Autorenkollektiv

Van der Waals Systems

Ab Initio Studies of the Interactions in Van der Waals Molecules

Ad van der Avoird, Paul E. S. Wormer, Fred Mulder and R. M. Berns

Van der Waals Systems: Molecular Orbitals, Physical Properties, Thermodynamics of Formation and Reactivity

Pavel Hobza and Rudolf Zahradnik

Intermolecular Interactions and Anesthesia: Infrared Spectroscopic Studies

Ginette Trudeau, Paul Dupuis, Camille Sandorfy, Jean-Max Dumas and Maurice Guérin

AKADEMIE-VERLAG · BERLIN

Autorenkollektiv

VAN DER WAALS SYSTEMS

AKADEMIE-VERLAG · BERLIN
1981

Die Originalausgabe erscheint im Springer-Verlag
Berlin · Heidelberg · New York
in der Schriftenreihe „Topics in Current Chemistry",
Volume 93

Vertrieb ausschließlich für alle Staaten mit Ausnahme der sozialistischen Länder:
Springer-Verlag Berlin · Heidelberg · New York
Vertrieb für die sozialistischen Länder:
Akademie-Verlag Berlin

Erschienen im Akademie-Verlag, 1080 Berlin, Leipziger Straße 3—4

Lizenznummer: 202 · 100/516/81
Gesamtherstellung: VEB Druckerei „Thomas Müntzer", 5820 Bad Langensalza
Umschlaggestaltung: Karl Salzbrunn
Bestellnummer: 762 939 1 (6618) · LSV 1215
Printed in GDR
DDR 68,— M

Springer-Verlag — Postfach 105280 · D-6900 Heidelberg 1
Telephone (06221) 487-1 · Telex 04-61723

Heidelberger Platz 3 · D-1000 Berlin 33
Telephone (030) 822001 · Telex 01-83319

Springer-Verlag New York Inc. — 175, Fifth Avenue · New York, NY 10010
Telephone 477-8200

Contents

Ab Initio Studies of the Interactions in Van der Waals Molecules

Ad van der Avoird, Paul E. S. Wormer, Fred Mulder, and Rut M. Berns

University of Nijmegen, Institute of Theoretical Chemistry, Toernooiveld, Nijmegen, Netherlands

Table of Contents

1 Introduction

Van der Waals molecules are complexes of molecules (or atoms) which are not held together by chemical bonding, as "normal" molecules, but by weaker Van der Waals forces. One of the main reasons to study experimentally the structure and spectra of Van der Waals molecules[1)] is to extract (rather detailed) information about the Van der Waals interactions between the constituent molecules, information which can be used for a better understanding and description of the properties of molecular gases, liquids and solids. For small molecules, up to about ten light atoms, even more detailed information about these Van der Waals interactions[1] can presently be obtained from ab initio calculations, i.e. directly from the approximate solution of Schrödinger's equation by variational or perturbational methods. The accuracy of the results is still a matter of concern, since the interactions are very small relative to the molecular total or (internal) binding energies and calculational errors which are larger than the Van der Waals binding energies are easily introduced. The collaboration between theory and experiment is here very useful: the experiment can serve as a check on the accuracy of the calculations. On the other hand, theoretical results can help in the interpretation of the spectra. Thus, experiment and theory can both be improved and the combined experience for small molecules can lead to physically justified, empirically parametrized model potentials for molecules larger than those for which the ab initio calculations are feasible.

Several books and review articles[2-13)] are concerned with Van der Waals interactions. In the present survey, we shall first describe which are the interaction mechanisms that hold Van der Waals molecules together and we shall concentrate on the dependence of these interactions on the orientations of the constituent molecules (at longer and shorter distances, sect. 2). Then, we outline some ab initio methods enabling reasonably accurate quantitative calculations of these interactions and we discuss possible sources of errors (sect. 3), some of which can be serious. Next, in section 4, we look at some results of ab initio calculations and their bearing on the structure and dynamics of Van der Waals molecules, after first dealing with some problems occurring in the representation of the ab initio results by analytical model potentials. The final section 5 describes the effect of intermolecular interactions on some properties of Van der Waals molecules other than the energy, and the quantitative calculation of these properties. Specifically, we discuss the interaction dipole moment and the interaction induced change in the polarizability, which are of importance for the intensities in infrared absorption and (inelastic) light scattering (Raman spectra); these are connected with the inter- and intramolecular vibrations in Van der Waals molecules.

We shall compare our (ab initio) results with experimental data, but we shall not deal in this paper with empirical or semiempirical determinations of Van der Waals interactions, since these are extensively described in the other surveys[1-13)].

---

[1] In many texts, the name Van der Waals interactions in reserved for the attractive long range forces between (neutral) molecules and, often, one only includes the leading R^{-6} term in the interaction energy; we use the name in a broader sense, meaning all the attractive and repulsive interactions between chemically non-bonding molecules (cf. sect. 2).

2 Mechanisms of Van der Waals Interactions; Distance and Orientational Dependence

The forces between the closed shell molecules in their electronic ground states (which are the constituents of most Van der Waals molecules studied at present) are of Coulombic origin. By this we mean that they originate from the Coulomb operator, describing the interaction between the electrons and nuclei in the complex. Even for (light) open-shell molecules, such as NO or O_2, the interactions between the magnetic spin and orbital momenta are expected to be smaller by several orders of magnitude than the electrostatic forces[14]. Relativistic (retardation) effects can be neglected for the distances of interest in Van der Waals molecules[14]. Therefore, the system of interacting molecules can be described by the time-independent non-relativistic Schrödinger equation. Practically always, when no electronic excitations or chemical reactions are considered, one can solve this Schrödinger equation in the Born-Oppenheimer approximation, i.e. one first obtains an effective potential for the nuclei by solving the equation for the electronic motion in the clamped nuclei approximation and then one calculates the nuclear (vibrational and rotational) states in this potential. Finally, in most work on intermolecular forces the rigid molecule approximation is made, i.e. it is assumed that the forces holding the nuclei together within one molecule are so much stronger than the intermolecular forces that the intramolecular and the intermolecular nuclear motions can be decoupled. Looking for example at ethylene (C_2H_4) molecules (sect. 4), the frequencies of the internal vibrations range from 3100 cm^{-1} (C—H stretch) to 810 cm^{-1} (out of plane bending), while we expect the C_2H_4—C_2H_4 vibrations in a Van der Waals molecule to lie below 150 cm^{-1}.

2.1 Distance and orientational dependence

Consider two rigid molecules A and B, both of arbitrary shape. Let $\vec{R} = (R, \underline{\Omega}) = (R, \Theta, \Phi)$ be the vector pointing from the center of mass of A to the center of mass of B. The coordinates of $\vec{R}$ are measured with respect to a space-fixed frame. Let the orientation of molecule A be described by the Euler angles $\underline{\omega}_A = (\alpha_A, \beta_A, \gamma_A)$, which are the angles associated with an (active) rotation of the molecule from an initial position in which a reference frame fixed on A is parallel to the space-fixed frame, to its present position. Similarly, the orientation of B is determined by the Euler angles $\underline{\omega}_B = (\alpha_B, \beta_B, \gamma_B)$. The interaction energy between A and B is most generally described by the following expansion[15,16]:

$$\Delta E^{AB}(\underline{\omega}_A, \underline{\omega}_B, \vec{R}) = \sum_{\Lambda} \Delta E_{\Lambda}(R)\, A_{\Lambda}(\underline{\omega}_A, \underline{\omega}_B, \underline{\Omega}) \tag{1a}$$

where $\Lambda \equiv (L_A, K_A, L_B, K_B, L)$ is the combination of "quantum numbers" applicable to the system A—B of interest, se table 1. The angular functions are defined as:

$$A_{\Lambda}(\underline{\omega}_A, \underline{\omega}_B, \underline{\Omega}) \equiv A_{L_A,K_A,L_B,K_B,L}(\underline{\omega}_A, \underline{\omega}_B, \underline{\Omega})$$

$$= \sum_{M_A,M_B,M} \begin{pmatrix} L_A & L_B & L \\ M_A & M_B & M \end{pmatrix} D^{L_A}_{M_A,K_A}(\underline{\omega}_A)^*\, D^{L_B}_{M_B,K_B}(\underline{\omega}_B)^*\, C^{L}_{M}(\underline{\Omega}). \tag{1b}$$

Table 1. Angles and quantum numbers specifying the orientational dependence of the interaction energy ΔE^{AB}, formula (1)

A	B	angular coordinates[a] $(\omega_A, \omega_B, \Omega)$	volume of angular coordinate space[a] V	quantum numbers[b] Λ
general molecule	general molecule	$\alpha_A, \beta_A, \gamma_A, \alpha_B, \beta_B, \gamma_B, \Theta, \Phi$	$256\pi^5$	L_A, K_A, L_B, K_B, L
linear[c] molecule	general molecule	$\alpha_A, \beta_A, \alpha_B, \beta_B, \gamma_B, \Theta, \Phi$	$128\pi^4$	$L_A, L_B, K_B, L (K_A = 0)$
atom	general molecule	$\alpha_B, \beta_B, \gamma_B, \Theta, \Phi$	$32\pi^3$	$L_B, K_B \quad (L_A = K_A = 0; L = L_B)$
linear[c] molecule	linear[c] molecule	$\alpha_A, \beta_A, \alpha_B, \beta_B, \Theta, \Phi$	$64\pi^3$	$L_A, L_B, L \quad (K_A = K_B = 0)$
atom	linear[c] molecule	$\alpha_B, \beta_B, \Theta, \Phi$	$16\pi^2$	$L_B \quad (L_A = K_A = K_B = 0; L = L_B)$
atom	atom	Θ, Φ	4π	$- \quad (L_A = K_A = L_B = K_B = L = 0)$

[a] One can choose a special coordinate system such that, for instance, $\Theta = \Phi = \alpha_B = 0$. So one needs a maximum of 5 (internal) angles in order to fix the (relative) orientations of the molecules in a dimer AB. This reduces the volume V by a factor of $8\pi^2$ (or 4π in the atom—atom case)

[b] From the behaviour of the angular functions (1b) under inversion of the total system, it follows that the summation (1a) over the quantum numbers Λ can be restricted to even values of $(L_A + L_B + L)$. If the molecules A or B have finite symmetry groups, Λ can be further restricted. For instance, if they have a center of inversion only terms with even L_A or L_B contribute. If A and B are identical molecules one can derive that:

$$\Delta E_{L_B, K_B, L_A, K_A, L} = (-1)^{L_A+L_B} \Delta E_{L_A, K_A, L_B, K_B, L} .$$

Relevant information can be found also in refs.[15] and[37]

[c] For linear molecules the remaining Euler angles can be chosen such that they coincide with the polar angles: $\alpha_A \equiv \phi_A, \beta_A \equiv \theta_A; \alpha_B \equiv \phi_B, \beta_B \equiv \theta_B$

The functions $\{D^{L_A}_{M_A, K_A}(\underline{\omega}_A); M_A = -L_A, \dots, L_A, K_A = -L_A, \dots, L_A\}$ constitute a $(2L_A + 1)$-dimensional matrix $\underline{\underline{D}}^{L_A}(\underline{\omega}_A)$ which represents the rotation $\underline{\omega}_A$ of molecule A. The set of these matrices forms a $(2L_A + 1)$-dimensional irreducible representation of the rotation group SO(3) [17]. In the active rotation convention, which we are using, the rotation matrices are given by[17, 18]:

$$D^{L_A}_{M_A, K_A}(\underline{\omega}_A) = e^{-i\alpha_A M_A} d^{L_A}_{M_A, K_A}(\beta_A) e^{-i\gamma_A K_A} \tag{2}$$

where $d^{L_A}_{M_A, K_A}(\beta_A)$ is a Wigner d-function[17]. The rotation matrices of molecule B are defined analogously. The symbol $\begin{pmatrix} L_A & L_B & L \\ M_A & M_B & M \end{pmatrix}$ stands for a 3-j coefficient and

$C^L_M(\underline{\Omega})$ is a Racah spherical harmonic, in the phase of Condon and Shortley, which can also be written as a special (K = 0) rotation matrix[17]:

$$C^L_M(\underline{\Omega}) = \left(\frac{4\pi}{2L+1}\right)^{1/2} Y^L_M(\Theta, \Phi) = D^L_{M,0}(\Phi, \Theta, 0). \tag{3}$$

This property (3) and the presence of the 3-j symbol in (1 b) makes the angular functions scalar, i.e. invariant under rotations of the total system (see Appendix). They span the complete space of scalar functions depending on $\underline{\omega}_A$, $\underline{\omega}_B$ and $\underline{\Omega}$ due to the completeness of the rotation matrices in the Hilbert space $L^2[SO(3)]$ (Peter-Weyl theorem[19]). The expansion, (1), is most convenient for molecular scattering calculations as well as for the solution of the nuclear motion problem in Van der Waals molecules (in terms of coupled translational and rotational vibrations of the rigid molecules), since it leads to a maximum separation of variables in the differential equations to be solved and it allows the power of angular momentum techniques to be employed.

For linear molecules A and B, where the interaction energy does not depend on the Euler angles γ_A and γ_B, only terms with $K_A = K_B = 0$ contribute and one can use (3) to obtain a simplified expression for (1)[20] (see also table 1):

$$\Delta E^{AB}(\theta_A, \phi_A, \theta_B, \phi_B, \vec{R}) = \sum_{L_A, L_B, L} \Delta E_{L_A, L_B, L}(R)\, A_{L_A, L_B, L}(\theta_A, \phi_A, \theta_B, \phi_B, \Theta, \Phi) \tag{4a}$$

with angular functions:

$$A_{L_A, L_B, L}(\theta_A, \phi_A, \theta_B, \phi_B, \Theta, \Phi) = \sum_{M_A, M_B, M} \begin{pmatrix} L_A & L_B & L \\ M_A & M_B & M \end{pmatrix} \times C^{L_A}_{M_A}(\theta_A, \phi_A)\, C^{L_B}_{M_B}(\theta_B, \phi_B)\, C^L_M(\Theta, \Phi). \tag{4b}$$

If one of the molecules, say A, is an atom in an S-state only the terms with $L_A = 0$ $(L = L_B)$ remain in (1) or (4). The same holds if we want to average over all orientations of one molecule, or, equivalently, put one molecule in the "unperturbed" rotational J = 0 state. When averaging over the orientations of both molecules, of course, only the isotropic contribution $\Delta E^{AB}_{\text{isotropic}} = \Delta E_{0,0,0,0,0}(R)$ remains.

In practical calculations of the intermolecular interaction potential one often chooses a special coordinate system with the z-axis parallel to $\vec{R}$ and such that $\alpha_B = 0$, which simplifies the angular functions (1 b) and (4 b), while still retaining all the dynamical coefficients $\Delta E_\Lambda(R)$. This simplification is easily introduced remembering that[17]:

$$C^L_M(0, 0) = \delta_{M,0} \quad \text{(Kronecker delta), independently of L.} \tag{5}$$

2.2 Model potentials

The dynamical coefficients $\Delta E_\Lambda(R)$, which are functions of the intermolecular distance only, fully determine the orientational dependence of the interaction potential. If one wishes to derive these functions from experimental data one has to replace them by relatively simple parametrized analytical or numerical forms. The simplest and most popular ones are:

a Lennard-Jones
n — 6 potential,
mostly with n = 12: $\Delta E_\Lambda(R) = A_\Lambda R^{-n} - B_\Lambda R^{-6}$ (6)

a Buckingham
exp —6(—8) potential: $\Delta E_\Lambda(R) = A_\Lambda \exp(-B_\Lambda R) - C_\Lambda R^{-6} - D_\Lambda R^{-8}$ (7)

but many other forms have been used, see the review by Pauly[21)].

In spite of the simple form of these distance functions and the usual assumption that the angular expansion can be truncated after very few terms (for instance, only the isotropic and the first anisotropic $L_A, L_B \neq 0$ terms), the number of parameters is mostly too large and these parameters are too strongly interdependent in affecting the measured properties, for a fully experimental determination of these parameters to be possible. Only for very simple systems such as atom-diatom systems[22-27)] or atom-tetrahedral molecule systems[28)] the experimental data could be used to yield a parametrized anisotropic potential of the form (1) and even there it appeared advantageous to extract part of the parameters[29,30)] from ab initio calculations. For other molecular systems only isotropic potentials[2] are known[21)], mostly in simplified forms such as (6) or (7).

Therefore, if one needs an anisotropic potential one often includes only specific anisotropic contributions, e.g. the molecular quadrupole-quadrupole interaction[32)], making the ad hoc assumption that all other anisotropic terms are small, or one invokes model potentials with fewer parameters which are intrinsically anisotropic. Examples of the latter are:

— the atom-atom potential, which assumes additive pair-wise isotropic interactions between the atoms p and q belonging to the different molecules (r_{pq} are the atom-atom distances):

$$\Delta E^{AB} = \sum_{p}^{\varepsilon A} \sum_{q}^{\varepsilon B} \Delta E(r_{pq}) \tag{8}$$

with $\Delta E(r_{pq})$ being, for instance, a Lennard-Jones potential (6) or a Buckingham potential (7) (with $\Lambda = 0$).

— elliptical scalings of isotropic molecule-molecule potentials[33-36)], for instance, a Buckingham exp — 6 potential (7) for $\Lambda = 0$ with the parameters A_0 and C_0 being simple function of the angles $\underline{\omega}_A$ and $\underline{\omega}_B$. (9)

The specific approximations which lie at the basis of these model potentials have to be justified, however, and they are not necessarily physically realistic. Actually, the truncated angular expansion of the potential (1), with parametrized functions $\Delta E_\Lambda(R)$ is a model potential also, which has to be verified.

Model potentials of the form (8) and (9) implicitly contain all the higher angular terms in (1) (with L_A, L_B up to infinity). Since the angular functions form an

[2] For atomic systems this is of course all one needs. Especially in the case of rare gases the (isotropic) potentials are known quite accurately[31)].

orthogonal set, one can explicitly calculate the dynamic coefficients ΔE_Λ for any known potential ΔE^{AB} by integrating over all angular coordinates:

$$\Delta E_\Lambda(R) = V^{-1}(2L_A + 1)(2L_B + 1)(2L + 1)\langle A_\Lambda(\underline{\omega}_A, \underline{\omega}_B, \underline{\Omega}) \mid \Delta E^{AB}(\underline{\omega}_A, \underline{\omega}_B, \vec{R})\rangle \tag{10}$$

with A_Λ being the angular expansion functions (1b) and V the total volume of the angular coordinate space, see table 1. In general, this integration must be carried out numerically; for specific model potential ΔE^{AB} analytical expressions have been derived[36–38], using angular momentum techniques.

2.3 Contributions to the Interaction Energy

Although the intermolecular forces which we consider here are all of electrostatic origin (in the broad sense used in the first paragraph of this section), we can distinguish different mechanisms which contribute to the interaction energy, and to other properties of Van der Waals molecules as well (see sect. 5). Let us denote the ground state electronic wave functions of the isolated (closed shell) molecules A and B by $|0^A\rangle$ and $|0^B\rangle$ and the corresponding "unperturbed" electron density distributions by ϱ_0^A and ϱ_0^B. The excited eigenfunctions of the molecular hamiltonians H^A and H^B we denote by $|a^A\rangle$ and $|b^B\rangle$ with corresponding eigenvalues E_a^A and E_b^B. For the interacting system we write the perturbation operator V^{AB} as:

$$\begin{aligned} V^{AB} &= H - H^A - H^B \\ &= \sum_i^{\varepsilon A} \sum_j^{\varepsilon B} \frac{Z_i Z_j}{r_{ij}}, \end{aligned} \tag{11}$$

the electrostatic interaction between all particles, electrons and nuclei, with charges Z_i and Z_j, belonging to A and B respectively.

2.3.1 Electrostatic; Long Range Multipole Interactions, Penetration Effects

The first order, electrostatic, interaction energy is defined as:

$$\Delta E^{(1)}_{elec.} = \langle 0^A 0^B | V^{AB} | 0^A 0^B\rangle = \int\int \varrho_0^A V^{AB} \varrho_0^B \, d\vec{r}_A \, d\vec{r}_B \tag{12}$$

which corresponds with the classical Coulomb interaction energy between the unperturbed molecular charge clouds. If these charge clouds do not overlap we can make a multipole expansion of this Coulomb energy or, equivalently, of the interaction operator V^{AB} [39–41]:

$$\begin{aligned} V^{AB} = &\sum_{l_A, l_B = 0}^{\infty} (-1)^{l_A} \left[\frac{(2l_A + 2l_B + 1)!}{(2l_A)!\,(2l_B)!}\right]^{1/2} R^{-l_A - l_B - 1} \\ &\times \sum_{m_A = -l_A}^{l_A} \sum_{m_B = -l_B}^{l_B} \sum_{m = -l_A - l_B}^{l_A + l_B} \begin{pmatrix} l_A & l_B & l_A + l_B \\ m_A & m_B & m \end{pmatrix} C_m^{l_A + l_B}(\Omega)\, \tilde{Q}_{m_A}^{l_A}(\vec{r}_A)\, \tilde{Q}_{m_B}^{l_B}(\vec{r}_B). \end{aligned} \tag{13}$$

The multipole operators are here defined as spherical tensors[17) 3]:

$$Q^{l_A}_{m_A}(\vec{r}_A) \equiv \sum_i^{\varepsilon_A} Z_i r^{l_A}_{i_A} C^{l_A}_{m_A}(\theta_{i_A}, \phi_{i_A}), \tag{14}$$

and analogously for B, in local coordinate systems of the molecules A and B respectively, which are parallel to the global space fixed coordinate system. Multipole moments are defined (and calculated) in a molecular (body fixed) frame, however, and therefore it is convenient to convert the multipole operators to this frame also. The frames of the molecules A and B are rotated by the angles $\underline{\omega}_A$ and $\underline{\omega}_B$ with respect to the space fixed frame and we can use the following property of the spherical multipole tensors (and the unitarity of the rotation matrices)[17)]

$$\begin{aligned}\tilde{Q}^{l_A}_{m_A} &= \sum_{m'_A} Q^{l_A}_{m'_A} D^{l_A}_{m'_A, m_A}(\underline{\omega}_A^{-1}) \\ &= \sum_{m'_A} Q^{l_A}_{m'_A} D^{l_A}_{m_A, m'_A}(\underline{\omega}_A)^* \end{aligned} \tag{15}$$

where $Q^{l_A}_{m'_A}$ are the multipole operators defined in the molecular frame of A and the $\tilde{Q}^{l_A}_{m_A}$ are the ones appearing in the expansion (13). The moments on B transform analogously. Substituting this relation into the multipole expansion (13) and this expansion into (12), the expression for $\Delta E^{(1)}_{elec.}$ immediately fits into the general expansion formula, (1). The dynamic coefficients in (1) obtain the closed form:

$$\begin{aligned}\Delta E^{(1)}_{\substack{mult.\\ L_A, K_A, L_B, K_B, L}} &= (-1)^{L_A}\left[\frac{(2L_A + 2L_B + 1)!}{(2L_A)!\,(2L_B)!}\right]^{1/2} \\ &\quad \times \delta_{L_A+L_B, L} R^{-L_A-L_B-1} \langle 0^A | Q^{L_A}_{K_A} | 0^A \rangle \langle 0^B | Q^{L_B}_{K_B} | 0^B \rangle, \end{aligned} \tag{16}$$

which represents the interaction between the K_A component of a permanent multipole moment (2^{L_A}-pole) on A and the K_B component of a permanent 2^{L_B}-pole on B, varying as $R^{-L_A-L_B-1}$ with distance.

For real molecules A and B the charge clouds have exponential tails so that there is always some overlap and the expansion (16) is an asymptotic series[43,44)]. Still, for the long range the multipole approximation to $\Delta E^{(1)}_{elec.}$ can be quite accurate, if properly truncated (for instance, after the smallest term). For shorter distances, the penetration between the molecular charge clouds becomes significant, the screening of the nuclei by the electrons becomes incomplete even for neutral molecules, and the power law for $\Delta E^{(1)}_{elec.}$ is modified by contributions which increase exponentially with decreasing R. These penetration contributions we define as:

$$\Delta E^{(1)}_{pen.} = \Delta E^{(1)}_{elec.} - \Delta E^{(1)}_{mult.}, \tag{17}$$

---

[3] For relations with Cartesian tensors, see ref.[42)].

with the "exact" electrostatic energy $\Delta E^{(1)}_{\text{elec.}}$ calculated according to (12) using the exact interaction operator (11), and $\Delta E^{(1)}_{\text{mult.}}$ obtained by summing an appropriate number of multipole interaction terms (16). A simple illustration of the occurrence of these penetration interactions is given by the example of two rare gas atoms, where $\Delta E^{(1)}_{\text{mult.}}$ is exactly equal to zero (term by term), while $\Delta E^{(1)}_{\text{elec.}}$ is not, if the charge clouds penetrate each other.

2.3.2 Induction, Dispersion; Multipole Interactions, Penetration Effects

The second order interaction energy, according to Rayleigh-Schrödinger perturbation theory is given by:

$$\Delta E^{(2)} = \sum_{a,b \neq 0,0} \frac{|\langle 0^A 0^B | V^{AB} | a^A b^B \rangle|^2}{E_0^A + E_0^B - E_a^A - E_b^B} . \tag{18}$$

Higher order terms can be defined as well, but what little experience is available has taught us that they are generally smaller by at least an order of magnitude. Still, they can be important if we look at specific effects such as the non-pairwise additive components to the interaction energy in Van der Waals trimers or multimers (cf. the last part of this section). In the second order summation over excited states (18) we can separate three different contributions:

$$\begin{aligned} \Delta E^{(2)} &= \sum_{a,b \neq 0,0} \cdots \\ &= \sum_{\substack{a=0 \\ b\neq 0}} \cdots + \sum_{\substack{a\neq 0 \\ b=0}} \cdots + \sum_{\substack{a\neq 0 \\ b\neq 0}} \cdots \\ &= \Delta E^{(2)}_{\text{ind. B}} + \Delta E^{(2)}_{\text{ind. A}} + \Delta E^{(2)}_{\text{disp.}} . \end{aligned} \tag{19}$$

It is easy to see that the first term corresponds with the classical polarization (or induction) energy of molecule B in the electric field of the electronic charge distribution ϱ_0^A plus the nuclei of A, the second term with the induction energy of molecule A in the field of ϱ_0^B plus the nuclei of B, while the third term, the dispersion energy, has no classical equivalent.

For the long range we may again substitute the multipole expansion (13) for V^{AB} and the rotation relation of the multipole operators (15), but, in contrast with the first order multipole interaction energy, the resulting expression does not immediately correspond with the general formula, (1). After recoupling the spherical tensors[41], the simple orientational dependence of (1) is recovered, however, and we find the following expressions for the dynamic coefficients[45]:

$$\begin{aligned} \Delta E^{(2)}_{\substack{\text{ind. A, mult.} \\ L_A, K_A, L_B, K_B, L}} = \sum_{l_A, l'_A, l_B, l'_B} & - 1/2\, \zeta^{L_A, L_B, L}_{l_A, l'_A, l_B, l'_B} R^{-l_A - l'_A - l_B - l'_B - 2} \\ & \times \alpha_{(l_A, l'_A) L_A, K_A} [\langle 0^B | \underline{Q}^{l_B} | 0^B \rangle \otimes \langle 0^B | \underline{Q}^{l'_B} | 0^B \rangle]^{L_B}_{K_B} , \end{aligned} \tag{20}$$

an analogous expression for $\Delta E^{(2)}_{ind.\,B}$ and:

$$\Delta E^{(2)}_{\substack{disp.,\,mult.\\ L_A, K_A, L_B, K_B, L}} = \sum_{l_A, l'_A, l_B, l'_B} - \zeta^{L_A, L_B, L}_{l_A, l'_A, l_B, l'_B} R^{-l_A-l'_A-l_B-l'_B-2}$$
$$\times \sum_{a \neq 0} \sum_{b \neq 0} (E^A_a - E^A_0 + E^B_b - E^B_0)^{-1}$$
$$\times [\langle 0^A | \underline{Q}^{l_A} | a^A \rangle \otimes \langle a^A | \underline{Q}^{l'_A} | 0^A \rangle]^{L_A}_{K_A}$$
$$\times [\langle 0^B | \underline{Q}^{l_B} | b^B \rangle \otimes \langle b^B | \underline{Q}^{l'_B} | 0^B \rangle]^{L_B}_{K_B} \tag{21}$$

The irreducible (spherical) multipole polarizabilities are defined as [41,42]:

$$\alpha_{(l_A, l'_A) L_A, K_A} = 2 \sum_{a \neq 0} (E^A_a - E^A_0)^{-1} [\langle 0^A | \underline{Q}^{l_A} | a^A \rangle \otimes \langle a^A | \underline{Q}^{l'_A} | 0^A \rangle]^{L_A}_{K_A}. \tag{22}$$

The symbol [.⊗.] stands for an irreducible tensor product[46] between two sets of tensors $\underline{T}^l = \{T^l_m; m = -l, \ldots, l\}$ and $\underline{T}^{l'} = \{T^{l'}_{m'}; m' = -l', \ldots, l'\}$:

$$[\underline{T}^l \otimes \underline{T}^{l'}]^L_M = \sum_{m, m'} T^l_m T^{l'}_{m'} (l, m, l', m' | L, M) \tag{23}$$

with $(l, m, l', m' | L, M)$ being a Clebsch-Gordan coefficient. The purely algebraic coefficient ζ occurring in (20) and (21) is lengthy but straightforward to calculate:

$$\zeta^{L_A, L_B, L}_{l_A, l'_A, l_B, l'_B} = (-1)^{l_A + l'_A} \left[\frac{(2l_A + 2l_B + 1)!\,(2l'_A + 2l'_B + 1)!}{(2l_A)!\,(2l'_A)!\,(2l_B)!\,(2l'_B)!} \right]^{1/2}$$
$$\times (2L_A + 1)^{1/2} (2L_B + 1)^{1/2} (2L + 1)^{1/2} (l_A + l_B, 0, l'_A + l'_B, 0 | L, 0)$$
$$\times \begin{Bmatrix} l_A & l'_A & L_A \\ l_B & l'_B & L_B \\ l_A + l_B & l'_A + l'_B & L \end{Bmatrix}, \tag{24}$$

the expression between curly brackets being a Wigner 9-j symbol[17]. For linear molecules all tensor components in (20), (21) and (22) with $K_A \neq 0$ and $K_B \neq 0$ are zero and, moreover, the permanent moments in (20) vanish for $m_B, m'_B \neq 0$ (in a body fixed frame with the z-axis along the molecular axis). The resulting formulas, which have been presented in ref.[20], are simpler[4].

---

[4] The expressions (16), (20) and (21) are easily related to the more standard form of the multipole expansion:

$$\Delta E^{(1,2)}_{mult.} = \sum_n C^{(1,2)}_n R^{-n}$$

which becomes, for anisotropic interactions:

$$\Delta E^{(1,2)}_{mult.}(\underline{\omega}_A, \underline{\omega}_B, \vec{R}) = \sum_n C^{(1,2)}_n(\underline{\omega}_A, \underline{\omega}_B, \underline{\Omega}) R^{-n}$$

by collecting all terms in the series (1) with $L_A + L_B + 1 = n$ in first order, (16), and all terms with $l_A + l'_A + l_B + l'_B + 2 = n$ in second order, (20) and (21). In the second order one can restrict the summations to $(l_A + l'_A + L_A)$ even and $(l_B + l'_B + L_B)$ even.

Again, for the short range, the second order energy contributions, (18), (19), calculated with the exact operator V^{AB} (11) start to deviate from the R power series expansions, (20) and (21), and we can define the penetration effects:

$$\Delta E^{(2)}_{\substack{\text{ind. pen.}\\ \text{disp. pen.}}} = \Delta E^{(2)}_{\substack{\text{ind.}\\ \text{disp.}}} - \Delta E^{(2)}_{\substack{\text{ind. mult.}\\ \text{disp. mult.}}} \tag{25}$$

which increase exponentially with decreasing R.

2.3.3 Exchange

Another effect, one which becomes dominant for the intermolecular forces at shorter distances, is the exchange effect, related to the required antisymmetry of the (exact) many-electron space-spin wave function under electron permutations (Pauli postulate, Fermi-Dirac statistics). In Rayleigh-Schrödinger perturbation theory, which works with product functions, $|a^A\rangle\,|b^B\rangle$, that are only antisymmetric with respect to the electron permutations within the subsystems A and B, this effect is not explicitly taken into account. Still, the total perturbation series, if it converges, will sum up to the exact wave function and the corresponding exact energy including exchange effects. Although it has been shown on simple model systems that this actually holds in practice[47)], convergence to the correct permutation symmetry is reached only in very high orders of perturbation theory. Moreover, there is the problem that the "exact" wave function of the system to which the Rayleigh-Schrödinger series starting with $|0^A\rangle\,|0^B\rangle$ converges, cannot obey the Pauli principle due to an incorrect symmetry in the spatial electron coordinates[44,48)]; so, it does not correspond to a physical state of the system (in case of more than two electrons).

Therefore, one would like to modify (symmetry adapt) the normal Rayleigh-Schrödinger perturbation theory such that the exchange effects are explicitly included in the lower order interaction energy expressions. This symmetry adaptation can be achieved by means of a projection operator, the antisymmetrizer A, that, operating on any N-electron space-spin function ($N = N^A + N^B$), makes this function antisymmetric under electron permutations. Furthermore, one must adapt the wave functions to the total spin operator. Using the relation between the electron spin functions, carrying representations of the group SU(2), and the irreducible representations of the permutation group S_N, projectors A_S can be defined[49)] that yield directly eigenfunctions of the total spin operator S^2 or their spin-free equivalents[50,51)]. Moreover, if the total interacting system AB contains spatial symmetry operations under which the products $|a^A\rangle\,|b^B\rangle$ are not invariant, A_S can be combined with operators that project the desired spatial symmetry as well.

It is not possible, however, to simply project the product functions $|a^A\rangle\,|b^B\rangle$ with A_S and then to use these functions in Rayleigh-Schrödinger perturbation theory, for two reasons. First, the projected functions $A_S|a^A\rangle\,|b^B\rangle$ are not eigenfunctions of the unperturbed hamiltonian $H_0 = H^A + H^B$ since H_0, which corresponds to a certain assignment of electrons to each subsystem A or B, does not commute with A_S. The total hamiltonian of the interacting system $H = H_0 + V^{AB}$ does commute with A_S, however, leading to the relation:

$$[A_S, H_0] = [V^{AB}, A_S] \neq 0 . \tag{26}$$

This relation shows how the action of the antisymmetrizer can mix different orders in perturbation theory. Secondly, the projected functions $A_S |0^A\rangle |0^B\rangle$ do not form an orthogonal set in the antisymmetric subspace of the Hilbert space $L^2(R^{3N})$; if we take all excited states $|a^A\rangle$ and $|b^B\rangle$ in order to obtain a complete set $|a^A\rangle |b^B\rangle$, the projections $A_S |a^A\rangle |b^B\rangle$ form a linearly dependent set. Expanding a given (antisymmetric) function in this overcomplete set is always possible, but the expansion coefficients are not uniquely defined. How the different symmetry adapted perturbation theories that have been formulated since the original treatment by Eisenschitz and London in 1930[52], actually deal with these two problems can be read in the following reviews:[53-56]. Usually, the first order interaction energy, including exchange effects, is defined by:

$$\Delta E^{(1)} = \frac{\langle 0^A 0^B | A_S V^{AB} | 0^A 0^B \rangle}{\langle 0^A 0^B | A_S | 0^A 0^B \rangle} \tag{27}$$

which, for two hydrogen atoms, corresponds with the Heitler-London interaction energy. The second and higher order energies (and the first and higher order wave functions) have different definitions in the different formalisms, however, which is related to the non-uniqueness problems mentioned above (non-uniqueness in the orders of perturbation and in the expansion coefficients of the perturbed wave functions).[5] To our opinion, preference to one or the other formalism should only be given on practical grounds: which perturbation expansion converges fastest, i.e. includes as much as possible the exchange contributions in the lower orders already; which expressions are easiest to evaluate. Theoretical and numerical comparisons can be found in the literature[54-58]. Actually, none of the second order exchange energies has been quantitatively calculated up to now for systems larger than two beryllium atoms[59]. The first order exchange energy can be defined as follows:

$$\Delta E^{(1)}_{exch.} = \Delta E^{(1)} - \Delta E^{(1)}_{elec.} \tag{28}$$

with $\Delta E^{(1)}$ and $\Delta E^{(1)}_{elec.}$ given by (27) and (12), respectively.

2.4 Interactions from Supermolecule Calculations

Because of the formal and practical problems with symmetry adapted perturbation theory, one mostly invokes variational methods applied to the total energy of the

[5] Physically, one can think of rather specific second order effects caused by the exchange "forces". For instance, the Pauli exchange repulsion between two closed shell systems leads to an outward polarization of the electron clouds which lowers this exchange repulsion. This energy lowering, which may be called exchange-induction energy, has indeed been found in variational calculations. The mathematical expression for this effect is not unique, however.

interacting system AB, thus performing so-called supermolecule calculations. The methods are usually the standard (ab initio) methods used also for calculating energies and wave functions of normal molecules[60], for instance the Hartree-Fock (SCF)—LCAO method possibly extended with Configuration Interaction (CI), Multi-Configuration (MC) SCF or Many-Body Perturbation Theory (MBPT). Since these methods work with fully antisymmetrized wave functions, all exchange effects are taken into account, but due to the fact that the total hamiltonian commutes with the antisymmetrizer A_S we avoid all the theoretical problems plaguing perturbation theory. The interaction energy is obtained by subtracting the subsystem energies:

$$\Delta E^{AB} = E^{AB} - E^{A} - E^{B} \tag{29}$$

and the first order interaction energy is usually defined as:

$$\Delta \tilde{E}^{(1)} = \frac{\langle A_S 0^A 0^B | H | A_S 0^A 0^B \rangle}{\langle A_S 0^A 0^B | A_S 0^A 0^B \rangle} - E^{A} - E^{B} \tag{30}$$

with $E^A = \langle 0^A | H^A | 0^A \rangle$, $E^B = \langle 0^B | H^B | 0^B \rangle$.

It is easily demonstrated, using the commutation relation (26) and the idempotency of the (hermitean) projector A_S, that this definition of $\Delta\tilde{E}^{(1)}$ is identical to the perturbational definition (27) of $\Delta E^{(1)}$ if the separate molecule wave functions $|0^A\rangle$ and $|0^B\rangle$ are exact eigenfunctions of H^A and H^B. For approximate molecular wave functions occurring in practical calculations, $\Delta E^{(1)}$ and $\Delta\tilde{E}^{(1)}$ are different, although for some approximations e.g. wave functions near the Hartree-Fock limit the deviations may be very small[61]. Also higher order interaction energies can be defined in a variational supermolecule treatment by making a perturbation expansion of the secular problem[62]. If the supermolecule wave function is expanded in terms of antisymmetrized products $A_S |a^A\rangle |b^B\rangle$, as in the multistructure Valence Bond method[63], the second (and higher) order energies become identical, for the long range, to the perturbational contributions (20), (21).

Such a correspondence cannot be found, if we use, for instance, the SCF—LCAO method with molecular orbitals delocalized over the entire supermolecule. Only indirect partitionings of the interaction energy are possible then, on the basis of different calculations with and without allowing delocalization and by component analyses of the wave function[64] (e.g. looking at the admixture of excited and/or ionic states to the "starting function" $A_S |0^A\rangle |0^B\rangle$). One must be cautious with such partitionings since they are basis set dependent. So, for instance, what has been called the charge transfer stabilization energy, or the charge resonance energy in case of identical subsystems A and B, is a second order overlap effect which can very well be interpreted[65] as the effect of charge penetration on the induction and dispersion interactions. Moreover, if the basis sets used in supermolecule calculations are too small, one finds contributions to the interaction energy which are artifacts of the calculation. An effect which is well known by now is the basis set superposition error[66], i.e. the energy lowering of each subsystem in a limited basis due to the admixture of basis functions centered on other subsystems. This is a purely mathematical effect, which automatically occurs in any variational supermolecule calculation where one allows electron delocalization. In some calculations a large part of the "charge transfer stabilization energy" should probably be ascribed to this artifact.

Another important (theoretical) point to note in supermolecule calculations is that the independent particle (SCF) model applied to the supermolecule AB includes all first order and the second order induction contributions to the interaction energy, correctly accounting for exchange, but not the second order dispersion contribution. The latter can be considered as an intermolecular electron correlation effect. In principle it can be obtained from supermolecule CI (or MCSCF, or MBPT) calculations, but in practice one must be very cautious again. Especially when using delocalized wave functions, the intermolecular correlation energy may easily get lost in the intramolecular correlation energy, which is typically hundred times larger. The basis set superposition error which occurs on the CI level[67,68] as well as on the SCF level[63,66] and which gives a distance dependent energy lowering, must be separated from the physical interaction contributions. Practical consequences of these problems will be discussed in the next section.

2.5 Additivity

The question how well the different contributions to the interaction energy in a composite system are additive can be asked on three different levels. First one may ask this question with respect to the different components (electrostatic, induction, dispersion, exchange) of the intermolecular (A—B) potential. If they were all calculated by standard (Rayleigh-Schrödinger) perturbation theory using the same interaction operator V^{AB} they would be exactly additive. Usually the "long range" contributions (electrostatic, induction, dispersion) are obtained from Rayleigh-Schrödinger perturbation theory, (mostly with the multipole expansion for V^{AB}), but the exchange effects are neglected then. In symmetry adapted perturbation theory the latter appear as modifications of the electrostatic, induction and dispersion energies; they can be additively separated by using definitions such as (28) for the first order exchange energy. In practice, the exchange contributions are mostly obtained from supermolecule calculations, together with electrostatic and induction energies in supermolecule SCF, or together with the electrostatic energy in $\Delta\tilde{E}^{(1)}$ (30). Then one can ask whether it is allowed to add dispersion (and induction) energies, calculated by second order perturbation theory. In other words, how large are the exchange-dispersion and exchange-induction energies? Are these not partially included in the supermolecule treatment already? And, if one uses the multipole expansion for the second order energies, how important is the additional neglect of the second order penetration energy (25)? The few data available for very small systems such as He_2[54,69,70] and Be_2[59] indicate that these second order exchange and penetration energies, in contrast with their first order counterparts (see section 3), are not very important (less than about 10% of the total second order energy) at the Van der Waals minimum.

The second additivity problem concerns the question whether the interaction potential in Van der Waals trimers or multimers (or molecular solids or liquids) is a sum of pairwise intermolecular (A—B) potentials. This question can be considered for each component of the interaction energy. The (first order) electrostatic energy is exactly pairwise additive. The dispersion energy has a three-body component which appears in third order of perturbation theory (the Axilrod-Teller triple-dipole interaction[5]), but which is only a few percent of the pair energy for distances of interest. The induction energy is not at all pairwise additive; the electric fields $\underline{F}_A$ and $\underline{F}_B$ originating from molecules A and B can be added, but the second order polarization energy of a third molecule C contains mixed (three body) terms of the type $-{}^1/_2 \underline{F}_A \underline{\underline{\alpha}}_C \underline{F}_B$ which can be of equal size as the quadratic terms. In many Van der Waals molecules the total induction energy is small, however, compared with the dispersion energy. The (relative) error introduced in the pairwise addition of exchange energies is of the order of the intermolecular overlap integrals; calculations on He_3, Ne_3 and Ar_3[71] show that it is small at the Van der Waals minimum. This intermolecular additivity problem is more extensively discussed by Margenau and Kestner[5], by Murrell[8] and by Claverie[72].

The third additivity question which is sometimes asked, regards the possibility of representing an intermolecular interaction potential as a sum of (isotropic) atom-atom potentials (8) (or bond-bond potentials). Not much is known about this question, since most of the atom-atom potentials used in practice are purely empirical. We consider this question in section 4 for C_2H_4—C_2H_4 and N_2—N_2 interactions on the basis of our information from ab initio calculations. Especially the exchange interaction can deviate from pairwise (atom-atom) additivity, which is not surprising as the intramolecular overlap integrals are considerable (of the order of 0.5, while the intermolecular overlap integrals are typically 0.01 at the Van der Waals minimum.

3 Quantitative Ab Initio Calculations

In this section we discuss the most important problems occurring when one wants to make quantitative calculations of the different interaction energy components. These components have been defined in the previous section for two molecules, denoted by A and B. We shall outline some practical methods for the calculation of pair interaction energies.

3.1 Methods

3.1.1 Molecular Wave Functions and Properties

The first requirement in both perturbational and variational calculations of the interaction energy, is the knowledge of the "unperturbed" molecular ground state energies, E_0^A and E_0^B, and the respective wave functions, $|0^A\rangle$ and $|0^B\rangle$. For many-electron systems A and B these are necessarily approximate. If the approximate wave functions $|0^A\rangle$ and $|0^B\rangle$ are to be used in calculating reasonably accurate first order interaction energies according to the expressions (12), (16), (27) or (30) they must be of sufficient quality to yield good multipole moments, $\langle 0^A|\, Q_{m_A}^{l_A}\, |0^A\rangle$ and

$\langle 0^B | Q^{l_B}_{m_B} | 0^B \rangle$; and, especially for the short range contributions, (17) and (28), the molecular charge distributions must be accurate in the intermolecular overlap region. For most (closed shell) systems reasonable accuracy (up to 10 or 20% error) in the interaction energies can be obtained with ab initio Hartree-Fock MO-LCAO or Roothaan wave functions[73] (single configuration functions, for closed shells single Slater determinants), provided one chooses good bases of atomic orbitals (AO's). These bases must be flexible especially in the outer regions of the molecules which determine the intermolecular overlap and which are weighted rather heavily in the multipole moments, particularly the higher ones. The same, or maybe an even more pronounced, sensitivity for the outer regions is exhibited also by the second order properties. Rules for selecting such bases are given in refs.[69,74-76]. The calculation of the required molecular electronic and nuclear interaction integrals and the solution of the Hartree-Fock LCAO equations can be performed routinely by any of the standard molecular SCF programs[60], usually based on Gaussian type AO's. The calculation of molecular wave functions and properties beyond the Hartree-Fock level (CI, MCSCF[60], MBPT[77]) is not a routine job yet, and the application of such wave functions to the evaluation of intermolecular interaction energies is even more difficult. The calculation of the molecular multipole moments can be carried out with the properties packages that go with some of the molecular wave function programs, up to octupole ($l = 3$) or hexadecapole ($l = 4$) moments, or with special programs for higher l values[78]. The lower multipole moments are sometimes available from experiment also.

3.1.2 Isotropic Long Range Interactions (Second Order)

In first order the isotropic long range interactions vanish, except when the molecules have charges (monopoles, $l = 0$). The next problem in a perturbation calculation of the interaction energy without intermolecular exchange is that in the second order energy, (18) or (21), and in the polarizability, (22), occurring in (20), one has to perform an infinite summation over the complete set of excited states $|a^A\rangle\, |b^B\rangle$. The majority of the work[6] on the estimation of this second order sum over excited states has been done in the multipole expansions, (20), (21), (22), where the matrix elements occurring in the numerator of $\Delta E^{(2)}$, (18), separate into products of monomer transition moments: $\langle 0^A | Q^{l_A}_{m_A} | a^A \rangle$ and $\langle 0^B | Q^{l_B}_{m_B} | b^B \rangle$. We summarize and illustrate the available ab initio methods for calculating the isotropic dispersion energy coefficients (and polarizabilities). These coefficients can be obtained from (21) by substituting $L_A = K_A = L_B = K_B = L = 0$ (cf. sect. 2):

$$\Delta E^{(2)}_{\substack{\text{disp., mult.}\\ \text{isotropic}}} = - \sum_{l_A, l_B} C_{\substack{2l_A+2l_B+2\\ \text{isotropic}}} R^{-2l_A-2l_B-2} \tag{31a}$$

and the multipole expansion coefficients are given by:

$$C_{\substack{2l_A+2l_B+2\\ \text{isotropic}}} = \frac{1}{4} \frac{(2l_A + 2l_B)!}{(2l_A)!\,(2l_B)!} \sum_{a \neq 0} \sum_{b \neq 0} (E^A_a - E^A_0 + E^B_b - E^B_0)^{-1} \times (E^A_a - E^A_0)^{-1} (E^B_b - E^B_0)^{-1} f^{0_A a_A}_{l_A} f^{0_B b_B}_{l_B} . \tag{31b}$$

Note that the mixed-pole terms ($l'_A \neq l_A$ or $l'_B \neq l_B$) have disappeared. The f_l^{0n} are the rotationally averaged 2^l-pole oscillator strenghts:

$$f_l^{0n} = 2(2l+1)^{-1} \sum_{m=-l}^{l} (E_n - E_0)\,|\langle 0|Q_m^l|n\rangle|^2 . \tag{32}$$

Let us introduce the moments of the oscillator strength distribution[79-81]. These are defined as:

$$S_l(k) = \sum_{n \neq 0} (E_n - E_0)^k f_l^{0n} . \tag{33}$$

The isotropic ($L_A = K_A = 0$) 2^{l_A}-pole polarizability of molecule A (cf. (22)) can be written as such a moment:

$$\underset{\text{isotropic}}{\alpha_{l_A}} = S_{l_A}(-2) . \tag{34}$$

The simplest (Unsöld) approximation[82] to the summations occurring in the expressions, (31), (33) and (34), is made by assuming that the excitation energies $(E_a^A - E_0^A)$ and $(E_b^B - E_0^B)$ can be replaced by constant average excitation energies Δ^A and Δ^B. In this manner, one obtains for the (Unsöld) polarizability (34):

$$\underset{\text{isotropic}}{\alpha_{l_A}^U} = S_{l_A}^U(-2) = (\Delta^A)^{-1}\, S_{l_A}(-1) \tag{35}$$

and one can avoid the summation over excited states $|n\rangle$ by using the closure relation (sum rule):

$$S_l(-1) = 2(2l+1)^{-1} \sum_m [\langle 0|Q_m^{l\dagger} Q_m^l|0\rangle - |\langle 0|Q_m^l|0\rangle|^2] . \tag{36}$$

The multipole coefficients in the dispersion energy (31) can now be written as:

$$\underset{\text{isotropic}}{C_{2l_A+2l_B}} = \frac{1}{4}\frac{(2l_A+2l_B)!}{(2l_A)!\,(2l_B)!}\frac{\Delta^A\Delta^B}{\Delta^A+\Delta^B}\; \underset{\text{isotrt}}{\alpha_{l_A}^U}\; \underset{\text{isotr.}}{\alpha_{l_B}^U} \tag{37}$$

which, for $l_A = l_B = 1$, is just London's well-known approximate expression for C_6[83]. Often, one assumes that the average excitation energies Δ can be replaced by empirical quantities, for instance the first molecular ionization energies. It is this latter assumption which has given the Unsöld approximation a rather bad reputation, since the results can be wrong by a factor of 2 (except for two hydrogen atoms where the error is less than 10%).

It is also possible to make ab initio calculations of the Δ's, however, and at the same time to improve the Unsöld scheme by assuming that the Δ's are dependent on the indices l, labelling the multipole operators (2^l-poles) associated with the excitations. Such a non-empirical Unsöld scheme has been proposed by Mulder et al.[75,84,85]. The average excitation energy is defined as the ratio:

$$\Delta_l = S_l(-1)/S_l(-2) \tag{38}$$

and the moments $S_l(k)$ are calculated for each molecule by considering all 2^l-pole excitations in a finite basis set. The Δ_l's are then substituted into (35) and (37) and the moments $S_l(-1)$ are computed from the sum rule expression (36), which holds exactly for exact wave functions. An advantage of this approach is that the ratio (38) appears to be rather insensitive to the quality of the basis set; the use of the sum rule (36) effectively corrects for the incompleteness of the basis.

Similar to this non-empirical Unsöld method is the generalized Kirkwood method[86-89] in its one-parameter version. Here Δ_l is defined as:

$$\Delta_l = S_l(0)/S_l(-1) \tag{39}$$

but the use of spectral representations (sums over excited states) is completely avoided by replacing both moments $S_l(0)$ and $S_l(-1)$ by their sum rule expressions, (36) for $S_l(-1)$. For example, in the dipole case we have the Thomas-Reiche-Kuhn sumrule[80] for $S_1(0)$:

$$S_1(0) = N \quad \text{(number of electrons in the molecule)} \tag{40}$$

Using the Hylleraas variation principle[80,90] with the (Kirkwood[91]) single parameter trial function $\lambda Q^l_m\,|0\rangle$ or with multi-parameter functions, it has been proved[86,89] that this method yields rigorous lower bounds to the dispersion multipole coefficients (31b) calculated from the (frequency dependent) polarizabilities via the Casimir-Polder formula[6,92]. This work, especially its recent extensions to larger sets of moments $S_l(k)$ yielding both upper and lower bounds to the dispersion coefficients[88], can be considered as an extension of the (semiempirical) methods by Langhoff, Gordon and Karplus[93], Pack et al.[94] and Meath et al.[95]. These authors have estimated dispersion energy coefficients from experimental oscillator strength distributions[95] and optical refractivity data[93,94]. This has yielded quite accurate results, but the experimental data required are only available for the dipole transitions, and so the application to higher moments must be based on ab initio calculations (cf. ref.[96] for rare gas dispersion interaction coefficients).

The most direct way to approximate the infinite summations occurring in (31), (33) and (34), is to replace them by finite summations over "effective excited states" [69,97,98]. In other words, the exact oscillator strength distributions of the molecules consisting of some delta functions for the discrete levels and a continuum for the ionized states, is replaced by a finite set of delta functions.[6] The completeness of these "effective excited states" can be tested by checking the sum rules for the moments $S_l(k)$ (33) and the Hylleraas variation principle can be invoked again to optimize the effective excited state wave functions (using a trial function which is a finite linear combination of such functions). Just as is the case for the ground state wave functions $|0\rangle$ one can in principle represent these excited states $|n\rangle$ at different levels of approximation, starting from Hartree-Fock LCAO (single configuration) wave functions and then correcting for the electron repulsion (correlation). The use of well-correlated wave functions (large CI expansions) becomes extremely difficult for molecules, however, if one has to describe all the excited states $|n\rangle$. Even at the Hartree-Fock LCAO level the calculations are already time-consuming, since one has to use large AO bases including atomic polarization functions in

---

[6] The pseudo-state method of Margoliash et al.[99] and the very simple Dalgarno refractivity scheme[100] for calculating molecular C_6 dispersion coefficients can be considered as empirical versions of this approach.

order to make the excited states $|n\rangle$ satisfy the sum rules to a reasonable accuracy[20,74,75]. The calculation of the molecular transition moments $\langle 0| Q^l_m |n\rangle$ is relatively easy[78].

3.1.3 Anisotropic Long Range Interactions

Some of the ab initio methods described in the previous section have been extended to the anisotropic interactions between molecules. For instance, the nonempirical Unsöld method, has been used[75,84,85] with (ab initio calculated) average excitation energies $\Delta_{l,m}$ that depend not only on the order l of the multipole operator associated with the excitations, but also on its components m. The method which explicitly calculates the "effective" excited state functions $|a^A\rangle$ and $|b^B\rangle$ has been applied[20,75,101] to the anisotropic long range interactions, expressions (20) and (21). An advantage of the latter method is that the functions $|a^A\rangle$ $|b^B\rangle$ can be used also in the "exact" formula for $\Delta E^{(2)}$, (18), where the multipole expansion is avoided, and, thus, charge penetration effects are included in $\Delta E^{(2)}$.[7] Even one can take (higher order) exchange effects into account in a variational treatment on the basis of the antisymmetrized states $A\,|a^A\rangle\,|b^B\rangle$[63].

Knowing the molecular permanent multipole moments and transition moments (or "closure moments" derived from sum rules, such as (36)), the computation of the first and second order interaction energies in the multipole expansion becomes very easy. One just substitutes all these multipole properties into the expressions (16), (20), (21) and (22), together with the algebraic coefficients (24) (tabulated up to R^{-10} terms inclusive in ref.[20], in a somewhat different form[8]), and one calculates the angular functions (1b) for given orientations of the molecules.

3.1.4 Intermediate Range Interactions

If one wants to calculate not only the long range part of the interaction potential, but also the region including the Van der Waals minimum, one must account for short range charge penetration and exchange effects. Charge penetration is included by evaluating the expressions (12) and (18) with the exact interaction operator V^{AB}, which requires the computation of very large numbers of electronic and nuclear interaction integrals. Although this computation can be performed with any of the standard integral programs for molecular calculations[60], it is a (computer) time-consuming job, especially as it has to be repeated for each dimer geometry. (In contrast with the multipole calculations where we have to repeat only the computation of the angular functions for each orientation of the molecules in the dimer.) Some time can be saved by modifying the standard programs so as to

---

[7] This could equally be done by making a spherical wave expansion for V^{AB}, as proposed by Koide[102], but his method for calculating $\Delta E^{(2)}$ is practical only for atoms.

[8] Although ref.[20] is concerned with linear molecules, this table is also complete for the general case of arbitrary molecules.

restrict the calculation to only those interaction integrals occurring in (12) and (18). Even this advantage is lost if one wishes to include exchange effects in the first, (27) or (30), or higher order interaction energies. The number of integrals becomes the same as in a full (supermolecule) calculation of the interacting system, AB. Moreover, one has to solve the supermolecule SCF, or CI, or MCSCF, or MBPT, or CEPA problem[60] or to calculate the different interaction contributions in (exchange) symmetry adapted perturbation theory[53–56]. To date, such calculations are possible only for very small molecules, that is, if one does not want to make severe restrictions on the size of the basis sets or to invoke approximate calculation schemes (e.g. the Extended Hückel[103], CNDO[104] or Gordon-Kim[105] method), which mostly lead to such a dramatic loss of accuracy that the calculated interaction potential looses all physical significance. Some typical computer timings for different steps in the calculation of the N_2—N_2 interaction potential are given in table 2. Before showing some illustrative results we discuss a few more points of practical importance.

Table 2. Typical computing times (CPU minutes) on IBM 370/158 (IBM 370/158 ≃ CDC 6400 ≃ Univac 1110) for the N_2—N_2 interaction potential surface[101,136]

	Interaction energy,[a] including penetration and exchange		Long range interaction energy[a] (multipole expansion up to R^{-10} terms inclusive)
Integrals	100–200	Multipole (transition) moments	3
$\Delta E^{(1)}$ (eqn. 30)	10	$\Delta E^{(1)}_{mult., A}$ (eqn. 16)	0.01
$\Delta E^{(2)}$ (eqn. 18), no exchange	15 (55)[b]	$\Delta E^{(2)}_{mult., A}$ (eqns. 20, 21)	15 (30)[b, c]
	125–225 min. Has to be repeated for each point on the potential surface		18 min. Potential surface can be generated by calculating $A_A(\omega_A, \omega_B, \Omega)$, (eqn. 1b)

[a] Monomer AO basis set of 72 primitive/50 contracted GTO's (9s, 5p, 2d/4s, 3p, 2d); integral time depends on distance R

[b] Monomer basis of 70 contracted GTO's ((basis a) + 1f)

[c] Reduced to 0.02 min. in the Unsöld approximation

In any (variational) supermolecule calculation which allows unconstrained mixing of basis functions centred on different molecules we obtain the so-called basis set superposition error (BSSE), the energy lowering of each subsystem by the addition of basis functions localized on other subsystems, (cf. sect. 2). This energy lowering is dependent on the dimer geometry and strongly resembles the shape of the physical interaction potential[63,67]; it is purely a numerical artifact, however, due to the fact that the monomer bases are not complete yet. The error occurs in principle at the SCF level[63,66] and, again, at the CI level[67,68,106]. One can reduce this error by using sufficiently large and well balanced orbital bases at the SCF level and configuration bases at the CI level, so as to reach the Hartree-Fock limit and the exact non-relativistic limit in the energies of the subsystems. At the SCF level this has now been done for dimers of small molecules (H_2), or atoms (He, Ne), up to the point where the BSSE is small compared with the interaction energy. At the CI level even with very large CI expansions for very small systems, such as He—H_2[107], the estimated error in the (ground state) correlation energy is still of the same magnitude as the Van der Waals well depth. At the SCF level one has proposed an approximate method[108] to correct for the BSSE. The energies of both monomers A, $E^{A(B)}$, and B, $E^{B(A)}$, are calculated in the AO basis of the total dimer AB. These energies are then subtracted from the dimer (SCF) energy E^{AB} in order to obtain the corrected supermolecule interaction energy (cf. expression (29)):

$$\Delta E^{AB} = E^{AB} - E^{A(B)} - E^{B(A)} \tag{41}$$

This procedure has been theoretically justified[109] and it appears to work numerically reasonably well[109,110] if the basis sets are not too badly chosen. At the CI level similar, though somewhat more complicated, procedures have been proposed[68,106,111], but there is less experience available yet as to their numerical performance.

A practical way to avoid contamination of the intermolecular correlation energy (which corresponds with the dispersion interaction) with geometry dependent intramolecular contributions which are numerical artifacts (for an analysis, see ref.[112]), is to express the CI wave functions in terms of orbitals localized on the monomers. By selecting the configurations one can then calculate the intermolecular correlation energy separately (in a CI or MCSCF or MBPT or CEPA calculation). For example, the multistructure VB method[63] is such a CI expansion with configurations $\mathcal{A}\,|a^A\rangle\,|b^B\rangle$ that are built from singly excited Hartree-Fock functions $|a^A\rangle$ and $|b^B\rangle$ and so they only contribute to the intermolecular correlation energy (according to Brillouin's theorem[113]). Actually, one should use correlated monomer wave functions, $|a^A\rangle$ and $|b^B\rangle$, or, in other words, include the coupling terms between intra- and intermolecular correlations. This has been done for very small systems: He—He[114,54], He—H_2[115], H_2—H_2[116]; it reduces the depth of the calculated Van der Waals well by about 10% in these systems. In other systems (Be, Mg, Ca atoms) where the monomer correlation energy is relatively large because of a quasi degenerate ground state and hence a single configuration wave function is a bad description, it has been found[117] that the effect of using correlated wave functions on the long range dispersion coefficients, C_6, C_8, C_{10}, can be considerable. It is important then[117] to "correlate" both the ground state $|0\rangle$ and the excited states $|n\rangle$ and the calculation of the full interaction potential becomes extremely difficult for all but the smallest systems.

3.2 Illustrative Results

We now show some representative results which illustrate the applicability of different methods for computing the intermolecular potential. We start with the long range part by looking at the first and second order multipole series (16), (20) and (21). The lowest term(s) in the first order series can be easily checked by comparing the (ab initio) calculated multipole moments (listed for many molecules in refs.[118,119]) with experimental ones that are available (usually only the dipole moment, sometimes the quadrupole and very rarely, such as for N_2[101], estimates for the next higher moment). For the leading second order dispersion term C_6R^{-6} (and for the dipole polarizability α_1 occurring in the leading induction term) reliable semi-empirical values are available (only the isotropic ones, mostly from refractivity data[93,94,100] and from optical spectra, yielding dipole oscillator strength distributions[95]). These can be used as a standard for the ab initio methods, while the latter have the great advantage that the higher coefficients and the anisotropy can be computed as well. The higher dispersion coefficients mostly quoted in the literature, so far, have been obtained from a single harmonic oscillator model [5,120,121]. This model has been reported[96,122,123] to underestimate C_8 and C_{10}, however, the more so for molecules of increasing size[124,125]. A more sophisticated

Table 3. Isotropic dispersion energy coefficients C_6 and ratios C_8/C_6 and C_{10}/C_6 (in atomic units) for various methods, described in the text

		Ab initio		Empirical[c]	Single-Δ Unsöld[d]	Oscillator model	
		Unsöld[a]	Kirkwood[b]			Fontana[e]	Amos and Yoffe[f]
C_6	NH_3	83.5	90.2	89.08		74	90.9
	CH_4	135	158	129.6		118	142 (133)
	N_2O	394	294	184.9		—	—
	C_2H_4	341	334	—		—	321 (357)
	C_3H_6	713	789	630.8		—	738
	C_6H_6	2670	2617	—		—	2300
C_8/C_6	NH_3	23.3	25.2		25.6 (19.0)	11.9	12.4
	CH_4	31.5	33.0		35.4 (23.3)	9.8	14.4 (14.1)
	N_2O	33.3	37.1		41.2 (25.3)	—	—
	C_2H_4	46.7	—		—	—	14.5 (17.5)
	C_3H_6	60.7	64.2		64.6 (33.0)	—	14.5
	C_6H_6	93.6	—		—	—	18.1
C_{10}/C_6	NH_3	672	692		726 (427)	174	188
	CH_4	1109	1123		1282 (626)	118	252 (241)
	N_2O	1369	1546		1776 (696)	—	—
	C_2H_4	2265	—		—	—	260 (398)
	C_3H_6	3589	3842		3905 (1211)	—	260
	C_6H_6	8345	—		—	—	428

[a] Results from refs.[84] for C_2H_4,[85] for C_6H_6 and[128a] for NH_3, CH_4, N_2O and C_3H_6

[b] Results from refs.[128a] for NH_3, CH_4, N_2O and C_3H_6,[128b] for C_2H_4 and C_6H_6 calculated according to the method of ref.[86]

[c] Accurate results from empirical dipole oscillator strength distributions for NH_3, N_2O and CH_4[95]; results for C_3H_6 according to the Dalgarno refractivity scheme[100] from ref.[128c]

[d] Results from ref.[128a] calculated with $\Delta_1 = \Delta_2 = \Delta_3$ (equal average energies for dipole, quadrupole, octupole excitations); values in parentheses if the original single-Δ Unsöld method is used as in ref.[123] with the sum rule (36) replaced by (42)

[e] Results from ref.[121]

[f] Results from ref.[126] for NH_3, CH_4 and C_2H_4; results in parentheses and those for C_3H_6 and C_6H_6 from ref.[129]. In both references the actual bond oscillator model values for C_8 and C_{10} were scaled upward by factors of 2 and 4, respectively; these scaled results are presented in the table

oscillator model has been introduced by Amos and Yoffe[126], who assigned different oscillators to different bonds in the molecule. A method to estimate ratios C_8/C_6 and C_{10}/C_6 first used by Margenau for H—H and He—He[127] has been revived by Starkschall and Gordon[123] in an application to rare gas atoms. This method is essentially identical with the Unsöld method mentioned above, but it assumes equal average excitation energies for the dipole, quadrupole and octupole transitions ($\Delta_1 = \Delta_2 = \Delta_3$). Recently, numerical values have been computed with this "single-Δ Unsöld" method for a number of molecular systems: CH_4, NH_3, H_2O, HF, (for ref. see[128]). In table 3 we compare the results obtained from these simplified models with results from the ab initio Unsöld and Kirkwood methods described above and with the accurate empirical data (for C_6 only). For C_6 the agreement is reasonable, except for N_2O (for more details, see ref.[128]). The ratios C_8/C_6 and C_{10}/C_6 are much smaller, however, in the oscillator model and in the single-Δ Unsöld method than in the ab initio Unsöld and Kirkwood schemes. These discrepancies are explained in ref.[128] as being largely due to further approximations made in the bond oscillator and single-Δ Unsöld models. In the original single-Δ Unsöld scheme[123] the sum rule expression (36) is replaced by

$$S_l(-1) = 2(2l+1)^{-1} \langle 0 | \sum_i r_i^{2l} | 0 \rangle, \tag{42}$$

thus neglecting significant terms. If these terms are included[128] the single-Δ Unsöld method yields ratios C_8/C_6 and C_{10}/C_6 which are slightly larger than the ab initio Unsöld values. In the bond oscillator model terms in C_8 and C_{10} are ignored which originate from the translation of the multipole operators from the molecular center of mass to the local bond origins. This causes the unexpected result, observed in refs.[126,129], that the ratios C_8/C_6 and C_{10}/C_6 hardly depend on the size of the molecules. When these translation terms are included the bond oscillator model gives somewhat more realistic C_8 and C_{10} values, the remaining discrepancies being due to the shortcomings of the harmonic oscillator model itself. The ratios C_8/C_6 and C_{10}/C_6 obtained with the ab initio Unsöld and Kirkwood methods are probably rather accurate; they might be more reliable even than the absolute C_6 values[128]. The insensitivity of the Unsöld method to the size of the basis (in contrast with the methods that explicitly include the excited states) is clearly displayed in fig. 9 of ref.[75]. Results calculated for H_2[20], N_2[101], C_2H_4[75,84], (aza)benzene(s)[84,85] by this method are quite satisfactory even for rather small bases. This figure and fig. 1 in this review also show the importance of the mixed-pole ($l_A \neq l'_A$ or $l_B \neq l'_B$) terms in the dispersion coefficients (21) higher than C_6; they cause the higher dispersion terms to be much more anisotropic than the first term[20,101].

After looking at the individual terms, we illustrate (on the example of two parallel ethene molecules) the convergence of the multipole series as a whole, in figs. 2 and 3, for the first order (electrostatic) and second order (dispersion) energies (16) and (21), and we compare the truncated multipole expansions with the unexpanded results, (12) and (18). The Van der Waals minimum in the isotropic potential lies at about R = 4.5 Å ($8.5a_0$). For large distance, R = 6.35 Å ($12a_0$), the multipole expansion accurately converges to the exact result, although the first

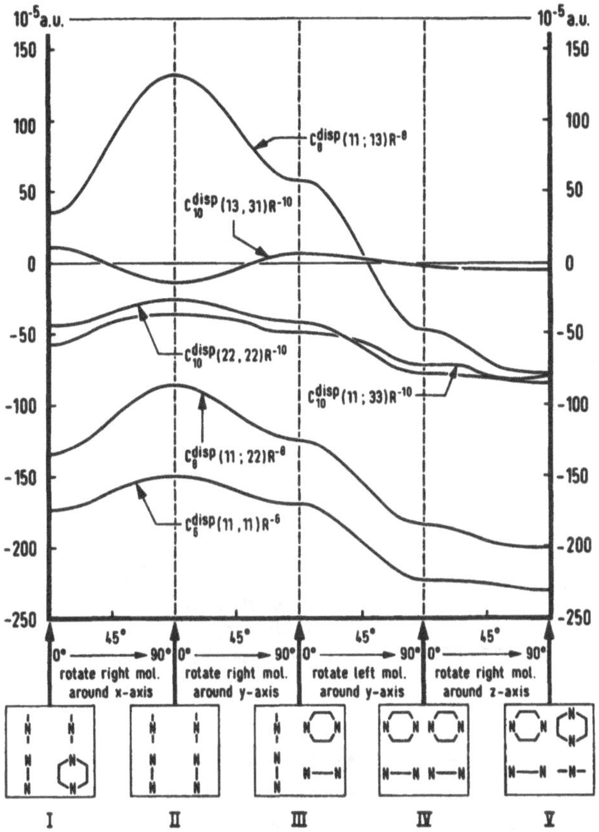

Fig. 1. Orientational dependence of the dispersion (multipole) interaction energy between two pyrazine molecules at $R = 10a_0$ (from ref.[84]). Different 2^l-pole contributions to eqn. (21) are labelled by $(l_A l'_A; l_B l'_B)$; quadratic terms: $l_A = l'_A$, $l_B = l'_B$; mixed pole terms: $l_A \neq l'_A$ or $l_B \neq l'_B$

term alone is in error by 25 and 10%, for the electrostatic and dispersion energies, respectively. At $R = 4.75$ Å ($9a_0$) the higher terms in the multipole series are even more important (the errors in the first term are 49 and 12%), but the series seems still convergent up to the point where we have truncated. At $R = 3.2$ Å ($6a_0$) the results clearly exhibit divergence of the series. Although the usual procedure of truncating the series after the smallest term and, possibly, a partial inclusion of this term[130–132] might give a crude estimate of the size of the interactions, it does not look very meaningful when the divergence starts already that early in the series. An alternative procedure of using damping functions[133] in order to correct the multipole series for charge penetration effects might work better, but one should realize that such damping functions probably have to be term and system dependent[134, 135]; then, they could only be obtained by actually evaluating the penetration effects and their use is not very helpful. We must warn for cases (a trivial one is the rare gas atom-atom interaction) where the multipole expansion does seem to converge, while the sum still

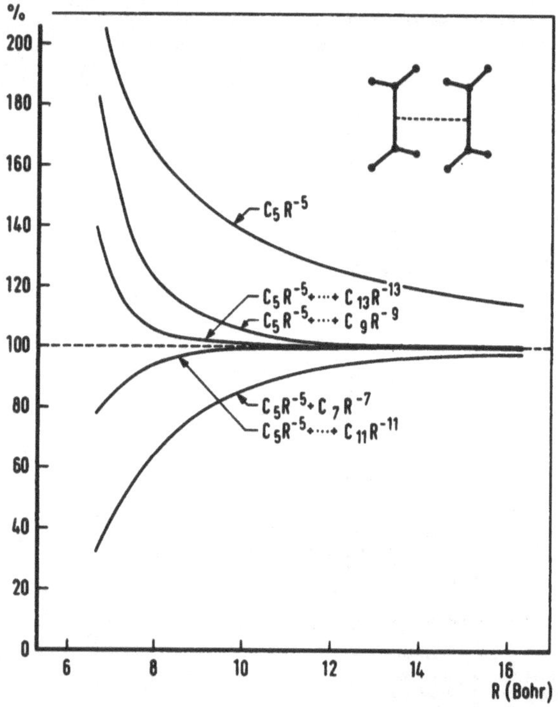

Fig. 2. Ratio of the multipole expanded (eqn. 16) and the unexpanded electrostatic energy (eqn. 12) for two parallel ethene molecules (from ref.[75]). Different multipole expansion lengths are shown

deviates from the unexpanded result, due to charge penetration. Such cases have been found for molecules also (see, for instance, fig. 4 for N_2—N_2).

If we look specifically at the anisotropic (L_A, $L_B \neq 0$) terms in the intermolecular potential, the convergence of the second order multipole expansion is slower than for the isotropic terms. This is caused by the strongly anisotropic mixed-pole ($l \neq l'$) contributions, which occur in the higher dispersion and induction multipole terms but not in the first term. This is illustrated in fig. 1 for the dispersion energy in the pyrazine dimer $(C_4N_2H_4)_2$. Fig. 5 shows that for the N_2 dimer the total anisotropy in the dispersion energy is comparable in size with the (purely anisotropic) electrostatic multipole interaction energy. The anisotropy is even stronger (relatively) in the induction energy, but the total induction energy is much smaller than the dispersion energy for the molecules we have considered: H_2, N_2, C_2H_4, benzene, azabenzenes, which have zero or small dipole moments.

After the long range interactions, we now consider explicitly the behaviour of the overlap (penetration, exchange) contributions to the interaction potential, particularly in the region around the Van der Waals minimum. In fig. 4 we have plotted these terms, together with the first and second order multipole interactions, as a function of distance for two parallel N_2 molecules, and in fig. 5 as a function of the molecular orientation at R = 4 Å. (The Van der Waals minimum in the isotropic N_2—N_2 potential lies at R = 4.1 Å[136]). The distance dependence is typical for closed shell molecules: with decreasing R we observe an attractive first order Coulomb interaction caused by charge penetration and a repulsive first order exchange interaction. Both increase exponentially, but the (Pauli) exchange repulsion dominates the

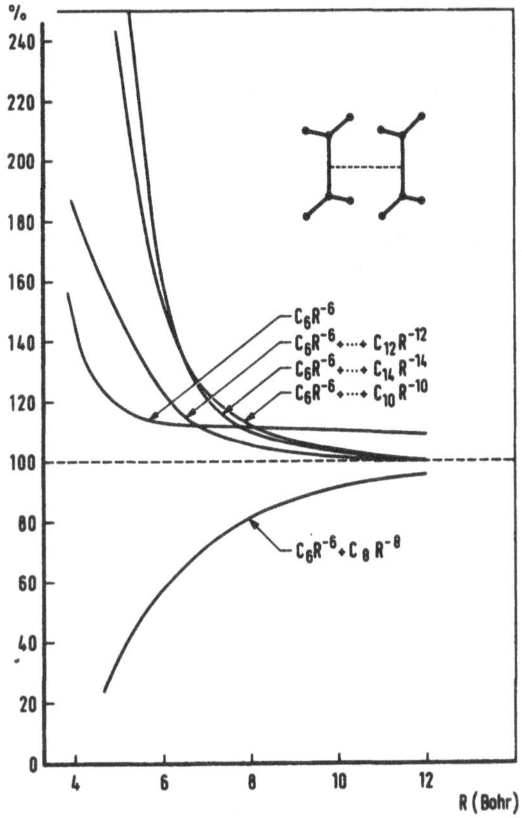

Fig. 3. Ratio of the multipole expanded (eqns. 20, 21) and the unexpanded dispersion and induction energy (eqn. 18) for two parallel ethene molecules (from ref.[75]). Different multipole expansion lengths are shown

penetration attraction by a factor of 5 to 10. The distance where these contributions begin to modify significantly the long range multipole interactions usually lies around the Van der Waals minimum: for N_2 it is slightly outside this minimum (which lies, for two parallel molecules at R = 3.6 Å), for H_2 it is slightly inside[137]. This depends on the diffuseness of the electron clouds, how far they protrude from the nuclear framework, as reflected, for instance, by the sign of the molecular quadrupole moment[138]. In the second order (dispersion and induction) energy the overlap effects are considerably smaller than in first order. In fig. 4 we see that they happen to be practically negligible for this particular (parallel) N_2—N_2 orientation; in other orientations they are somewhat larger. In fig. 5 we observe that the overlap contributions, although always repulsive, are the dominating anisotropic terms at distances around the Van der Waals minimum (or shorter). This will be reflected in the structure of the Van der Waals molecules (cf. sect. 4).

An interesting subject we may comment upon is the interaction between (planar) molecules with conjugated π-electron system (e.g. aromatic molecules). It is sometimes argued that these molecules show a particularly strong dispersion attraction because of the large in-plane polarizability of the π-electrons. In a series of ab initio calculations for benzene and several azabenzenes[84,85], Mulder et al.

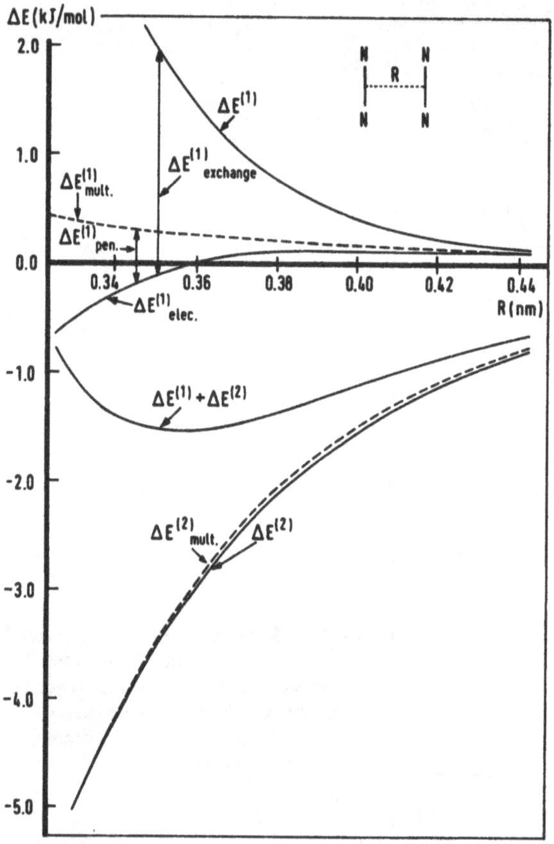

Fig. 4. Different contributions to the interaction energy between two parallel N_2 molecules; ab initio results from refs[101, 136].
$\Delta E^{(1)}_{elec.}$, eqn. (12),
$\Delta E^{(1)}_{mult.}$, eqn. (16), sum of complete R^{-n} terms for n = 5, 7, 9
$\Delta E^{(1)}$, eqn. (30)
$\Delta E^{(2)} \equiv \Delta E^{(2)}_{elec.}$, eqn. (18)
$\Delta E^{(2)}_{mult.}$, eqn. (21), sum of complete R^{-n} terms for n = 6, 8, 10
Short range penetration, $\Delta E^{(1)}_{pen.}$ eqn. (17), and exchange, $\Delta E^{(1)}_{exch.}$ eqn. (28), effects become visibly important with decreasing R (in first order)

have found however, (in contrast with earlier estimates[139, 140] and semi-empirical calculations[141]) that also the perpendicular π-polarizability is far from negligible. Moreover, it appeared that for none of the polarizability components the π-electron contribution really dominates over the contribution from the σ-skeleton. Also in determining the long range dispersion coefficients the π-electrons are certainly not dominant (less than 23% of C_6) for this class of molecules[85]. For larger molecules (naphthalene, anthracene, etc.) the relative π-contribution is expected to be somewhat larger, though[85].

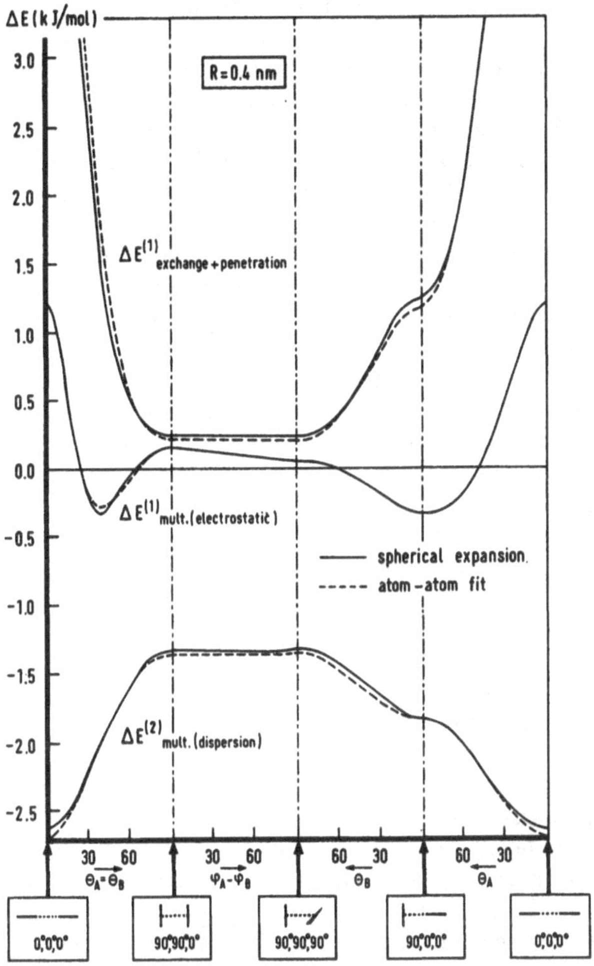

Fig. 5. Orientational dependence of different long range (multipole) and short range (exchange + penetration) contributions to the N_2-N_2 interaction energy, at R = 4Å, from ref.[136]. Orientations are described by the angles θ_A, θ_B, $\phi_A - \phi_B$, see table 1. Curves generated by the spherical expansion(4) and by the atom-atom fit (8) of the "ab initio" potential are shown

4 Structure of Van der Waals Molecules

If we want to predict the structure, the stability and the vibrational and rotational spectra of Van der Waals molecules, we have to know the complete intermolecular potential as a function of the intermolecular distance(s) and the molecular orientations. For rare gas dimers[31] and for some rare gas atom-diatomic molecule (e.g. H_2, HCl) systems rather detailed information about the potential is available from experiment[1, 22–27, 142, 143], from ab initio calculations[115, 124, 125, 144, 145] or both[29, 30].

The latter systems have only two internal degrees of freedom, however, (in the rigid molecule approximation) and the rare gas dimers have just a single one, of course. Some ab initio studies have been made of molecular Van der Waals (or hydrogen bonded) systems with more internal coordinates[119], but mostly they concern only specific points or one-dimensional cuts (e.g. distance curves for fixed molecular orientations) of the potential (hyper) surface. One exception is the case of the simplest molecular dimer $(H_2)_2$, which has been studied in detail, both ab initio[116, 124, 125, 146, 147] and experimentally[26, 148–154]. Another exception form the two Van der Waals molecules, $(C_2H_4)_2$ and $(N_2)_2$, of which the complete potential surfaces have been obtained in our institute[63, 75, 155, 101, 136] via ab initio calculations. The N_2—N_2 potential, in particular, has been the subject of much previous (semi-) empirical work[156]. The dimers $(N_2)_2$ [158, 159] and $(C_2H_4)_2$ [160, 161] have been investigated experimentally too, but even for $(N_2)_2$ where the IR spectrum is known[158], the structure could not be inferred from the experiments yet. Although we have not carried out the second step in the Born-Oppenheimer scheme, the solution of the nuclear motion, we shall, on the basis of our calculated potentials, make some remarks about the equilibrium structure, the binding and the internal molecular mobility in the Van der Waals molecules $(N_2)_2$ and $(C_2H_4)_2$. These remarks may be confronted with new experiments which are certainly to be expected in the near future.

4.1 Analytical Representation of the Intermolecular Potential; Fitting of the Ab Initio Results; Atom—Atom Potentials

For all but the very smallest systems, (such as HeH_2^+ [162] and even there it is very expensive), it is not possible in practice to calculate the full potential surface, with a grid fine enough that it can be directly used for solving the (nuclear) dynamical problem in Van der Waals molecules (or for scattering calculations). Moreover, such a numerical potential would not be convenient for most purposes. Therefore, one usually represents the potential by some analytical form, for instance, a truncated spherical expansion (1) or another type of model potential (cf. sect. 2). The parameters in this model potential can be obtained by fitting the ab initio results for a limited set of intermolecular distances and molecular orientations. Since we have encountered some difficulties in this fitting procedure which we expect to be typical, we shall describe our experience with the $(C_2H_4)_2$ and $(N_2)_2$ cases in some detail. At the same time, we use the opportunity to make a few comments about the convergence of the spherical expansion used for $(N_2)_2$ and about the validity of the atom-atom model potential applied to both $(C_2H_4)_2$ and $(N_2)_2$.

4.1.1 $(C_2H_4)_2$

For this dimer the interaction potential has been calculated[155, 75] for 8 different orientations of the two molecules, for 3 distances around $R = 4.8$ Å $(9a_0)$ including (first order) exchange and penetration effects and for 8 distances from $R = 6.4$ to 10.6 Å (12 to $20a_0$) in the multipole expansion. Second order overlap effects (penetration and exchange) were neglected. The number of orientations (8) is far too small (the system has 6 internal coordinates, 1 distance and 5 angles, and it needs all 5 quantum numbers, L_A, K_A, L_B, K_B, L, to qualify its angular functions (1 b)) to make a spherical expansion (1) of the interaction potential, even if one assumes that this expansion could be truncated after L_A, $L_B = 2$ terms. Some simplified model potential had to be adopted in order to reduce the number of fitting parameters. Wasiutynski et al.[155] have chosen an atom-atom potential (of the exp-6-1 type):

1. because it has relatively few parameters,
2. because it is easy to use in lattice dynamics calculations for the molecular crystal, and
3. because there is a considerable amount of work on hydrocarbons based on empirically parametrized atom-atom potentials[163-166] (from crystal heats of sublimation and structural data). The ab initio results could be used to check both the atom-atom model and its empirical parametrization.

At first, it was tried to optimize all the atom-atom parameters simultaneously by a best fit to the total ab initio interaction energies, but this procedure led to highly correlated fit parameters with no well-defined optimum. Then, the ab initio interaction energy was separated and three independent fits were made:

(i) the electrostatic (multipole) interaction energy (16) calculated up to R^{-7} terms inclusive was fitted by an atomic point charge model. If the point charges are fixed on the nuclei (which leads to a single independent charge parameter for the C_2H_4 molecule) the fit is bad (root mean square deviation 23%) especially for some C_2H_4 orientations. If we extend the model to 4 parameters by allowing the charges to shift away from the nuclei the fit is much better (r.m.s.d. 3.6%).

(ii) the dispersion (multipole) energy (21) truncated after R^{-8} terms was fitted by an r^{-6} atom-atom potential. This went quite well (surprisingly well, if we consider the incorrect asymptotic angular behaviour of the r^{-6} atom-atom potential[37,167]), but an averaging constraint had to be imposed on the C—H parameter, in order to avoid high correlation. The final fit, with only two independent parameters, had a r.m.s.d. of 7.1%. The induction energy, which is very small relative to the dispersion energy, was neglected.

(iii) the overlap (first order penetration and exchange) energy, (17) and (28) calculated from (30), was fitted by an exponential atom-atom potential. The electrostatic penetration energy was separated from the electrostatic multipole energy (i), since the atom-atom (point charge) model cannot acount for penetration effects. It was added to the exchange energy which has about the same exponential distance dependence. It is this distance dependence, which was actually found in ab initio calculations[145,155,168], that justifies the use of an exponential atom-atom repulsion, rather than an r^{-n} type. Just as in (ii) one had to put averaging contraints on the C—H parameters and, moreover, the H—H repulsion parameters had to be determined by comparing specific dimer geometries where the energy differences are mainly caused by H—H contacts. The final fit (with 4 independent parameters) is still rather unsatisfactory (r.m.s.d. 33%).

The different contributions to the interaction potential are displayed in figs 6, 7 and 8, for 8 different orientations of the C_2H_4 molecules in the dimer. These figures clearly illustrate the quality of the atom-atom model in representing the orientational de-

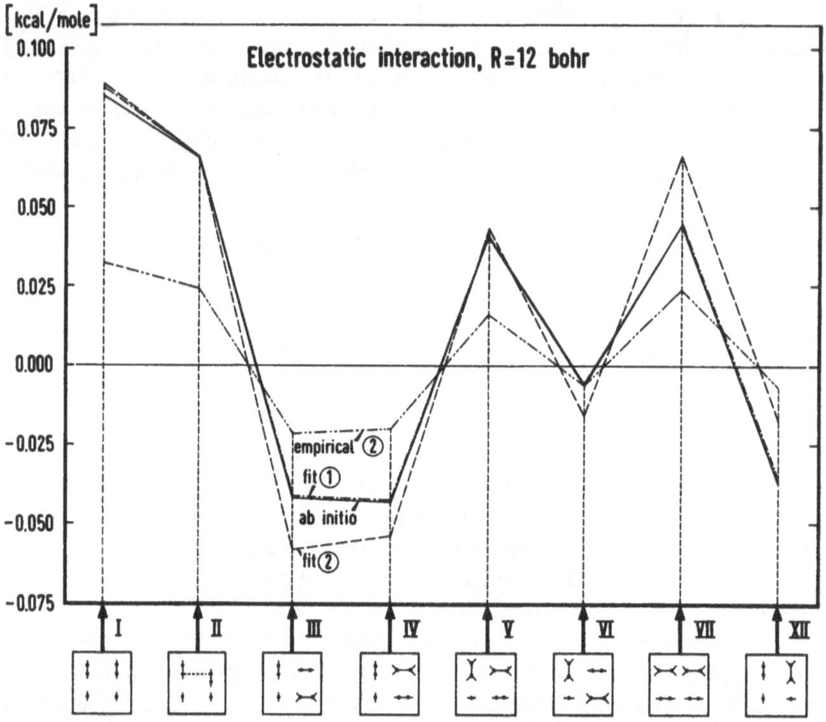

Fig. 6. Electrostatic multipole expanded interaction energy between two ethene molecules for 8 different orientations at R = $12a_0$ (ab initio results from refs.[75,155]). Fits ① and ② are atom-atom (point charge) model fits to the ab initio data; ① is a fit with the charges shifted from the nuclei, ② has the charges fixed on the nuclei. Also indicated is an empirical atom-atom potential; for details see ref.[155]

pendence of the interaction potentials; they also contain some empirical atom-atom potentials for comparison.

We can draw the following conclusions. For the dispersion (multipole) interaction energies the atom-atom model works reasonably in the $(C_2H_4)_2$ case, (and in other cases, too: benzene, azabenzenes[167]). Also the electrostatic (multipole) energy can be well represented by an atom-atom (point charge) model, if the point charges are allowed to shift away from the nuclei, (or if one adds extra point charges[169]). For the overlap energy the deviations from the ab initio results are much stronger. This might be due to deviations from pairwise additivity (cf. sect. (2), the intramolecular overlap between the atomic orbitals is considerable, $\simeq 0.5$). But also it can be caused partly by the anisotropy of the atom-atom interactions. Both effects are related to the chemical bonding within the molecules which is ignored in the atom-atom model. Actually, one can observe from the ab initio results[155] that the C—C overlap repulsion has a longer range (corresponding to a smaller negative exponent) in the direction perpendicular to the C_2H_4 plane than in the other directions. This is due to the relative diffuseness of the π-clouds. Fortunately, a deviation of 33% in the overlap energy does not have such a drastic effect on the potential surface as it may

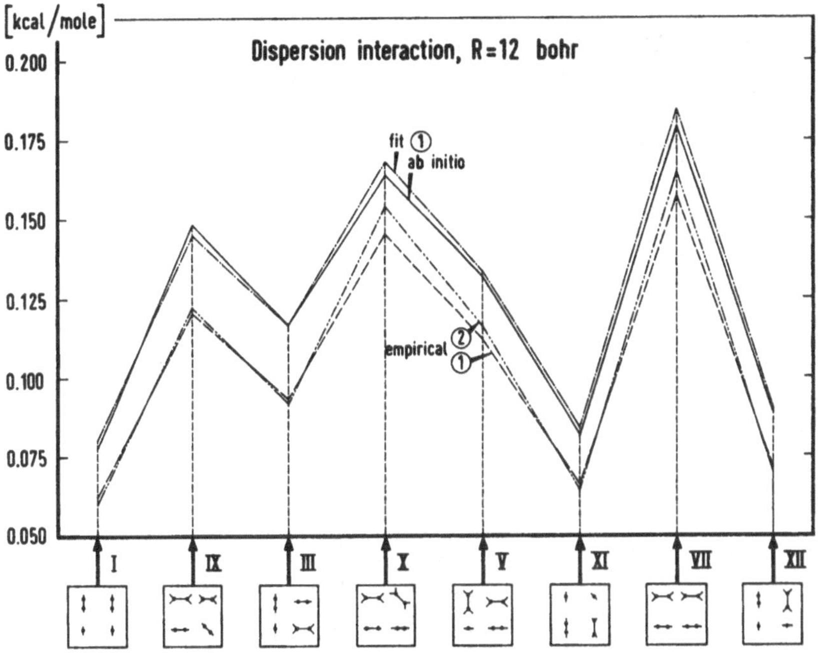

Fig. 7. Dispersion multipole interaction energy between two ethene molecules for 8 different orientations at R = $12a_0$ (ab initio results from refs.[75, 155]). Fit ① is an atom-atom fit to the ab initio data. Also indicated are two empirical atom-atom potentials; for details see ref.[155]

seem. The overlap repulsion depends very steeply on the intermolecular distance. A large deviation in the repulsive energy will cause a much smaller change in the equilibrium distance and not too much of a change in the depth of the Van der Waals well.

The atom-atom potential fitted to the ab initio data gives fairly realistic results[155] for the equilibrium structure[170] (unit cell parameters and molecular orientations in the cell), the cohesion energy[171] and the phonon frequencies of the C_2H_4 molecular crystal. The latter have been obtained via both a harmonic and a self-consistent phonon lattice dynamics calculation[155, 172] and they were compared with IR[173] and Raman[174] spectra. About some of the empirical hydrocarbon atom-atom potentials[164], which are fitted to the crystal data, we can say that they correspond reasonably well with the ab initio results (see figs. 6, 7, 8), their main defect being an underestimate of the electrostatic multipole-multipole interactions.

4.1.2 $(N_2)_2$

In this system the long range (multipole) interaction energy has been calculated[101] directly in the form of a spherical expansion (4): electrostatic R^{-5}, R^{-7} and R^{-9} terms, formula (16), dispersion R^{-6}, R^{-8}, R^{-10} terms, formula (21) and induction R^{-8}, R^{-10} terms, formula (20). The multipole moments used in the electrostatic energy

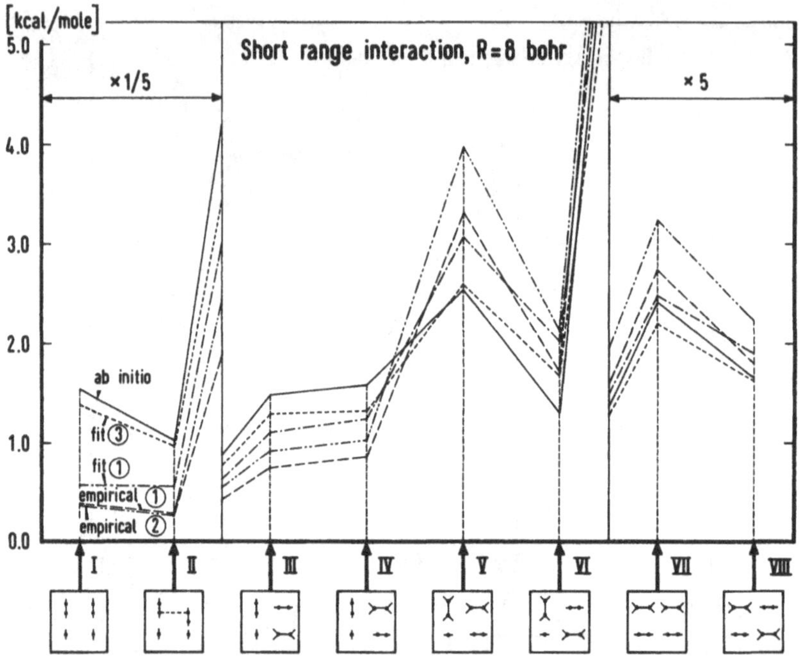

Fig. 8. Short range (exchange + penetration) interaction energy between two ethene molecules for 8 different orientations at $R = 8a_0$ (ab initio results from ref.[155]). Fits ① and ③ are atom—atom fits to the ab initio data, with and without constraints for the C—H parameters. Also indicated are two empirical atom—atom potentials; for details see ref.[155]

agree rather well with the experimental data available, but the calculated dispersion coefficient C_6 is considerably larger (25 %) than the accurate semi-empirical value[95]. Also the anisotropy in C_6 and in the dipole polarizability α are somewhat overestimated with respect to experiment. We believe this to be due to the use of Hartree-Fock wave functions for the N_2 monomers (the AO basis was sufficiently large and flexible). Employing the accurate semi-empirical data for C_6 and α in combination with the ab initio results, better estimates are given for the dispersion coefficients C_6, C_8, C_{10} and their anisotropic components[101]. We shall refer to these values as "ab initio". The induction energy is very small and can be neglected relative to the dispersion energy just as for $(C_2H_4)_2$.

In first instance, the overlap (penetration and exchange) energy has been computed (in first order, from expression (30)) for 6 different orientations of the two N_2 molecules and 5 distances[136]. Also the second order penetration contribution (25) was computed (for 10 of the 30 dimer geometries), but this contribution is small relative to the first order penetration contribution and it has been neglected. In principle, these ab initio data should be sufficient to calculate (for each R) 6 coefficients $\Delta E_{L_A, L_B, L}(R)$ in the spherical expansion (4). Alternatively one can obtain all (independent) terms up to L_A, $L_B = 2$, i.e. $(L_A, L_B, L) = (0, 0, 0)$, $(2, 0, 2)$, $(2, 2, 0)$, $(2, 2, 2)$, $(2, 2, 4)$, from the data for 5 orientations and use the 6th orientation for a check on the truncation error. The results of this procedure were completely unsatisfactory, however. Berns et al.[136] have found that the procedure of calculating a certain number of spherical expansion coefficients $\Delta E_{L_A, L_B, L}$ from the interaction energies ΔE^{AB} computed for an equal number of orientations ($\underline{\omega}_A$, $\underline{\omega}_B$, $\underline{\Omega}$) is numerically not very stable. Instead, they have proceeded, after some experimentation, by calculating the expansion coefficients of the (ab initio) overlap energy directly from expression (10), using a suitably chosen numerical integration procedure over the angular coordinates $\underline{\omega}_A$, $\underline{\omega}_B$, $\underline{\Omega}$ (θ_A, θ_B and ϕ_A are sufficient in this case). This involved the ab initio calculation of the first order energy (30) for 105 different orientations of the two N_2 molecules, in order to obtain the first 18 (independent) dynamic coefficients in the expansion (4). This was done for one distance, R = 3 Å, well inside the Van der Waals minimum (at R = 4.1 Å) of the isotropic potential. It was found that the coefficients $\Delta E_{L_A, L_B, L}$ indeed decrease with increasing L_A, L_B; for fixed L_A, L_B they increase with increasing L. Some of the highest coefficients calculated (for L_A, L_B = 4,4 and 6,2) were less than 1% of the isotropic coefficient $\Delta E_{0,0,0}$. It can be concluded that the series (4) converges, but that some of the higher terms are still important. Truncation of the series after L_A, L_B = 2,2 leads to an error of 16%, truncation after L_A, L_B = 4,4 to 2% error. Next, it was decided, on the basis of the ab initio results for 6 distances and 6 orientations, to represent all the coefficients $\Delta E_{L_A, L_B, L}(R)$ up to L_A, L_B = 6,2 by the same exponential function of R. This caused a somewhat larger error (7%), but it is certainly not a bad (first) approximation. The results, in combination with the long range results, yield a reasonable fit of the calculated N_2—N_2 interaction potential (see figs. 9 and 10).

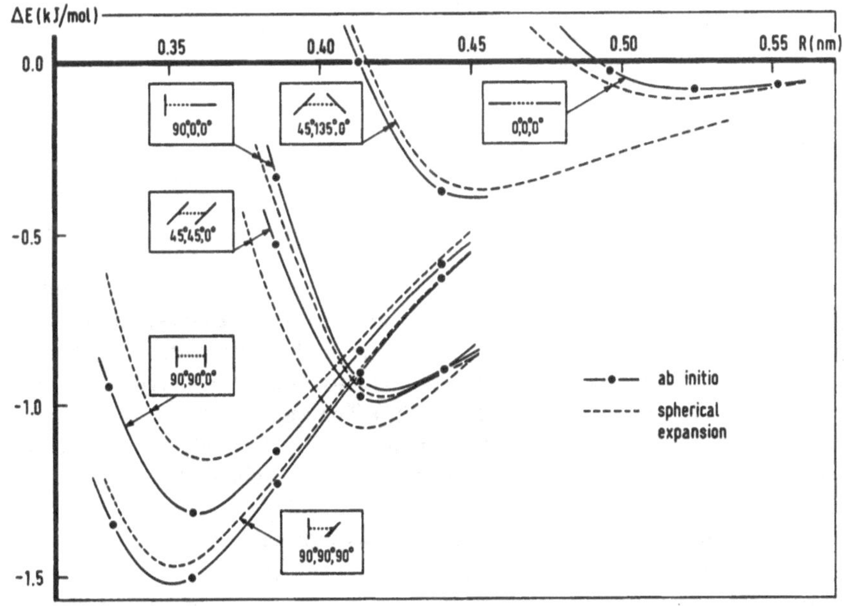

Fig. 9. Total interaction energy between two N_2 molecules at 6 different orientations, described by the angles θ_A, θ_B, ϕ_A. "Ab initio" results and spherical expansion (4) of these results from ref.[136]

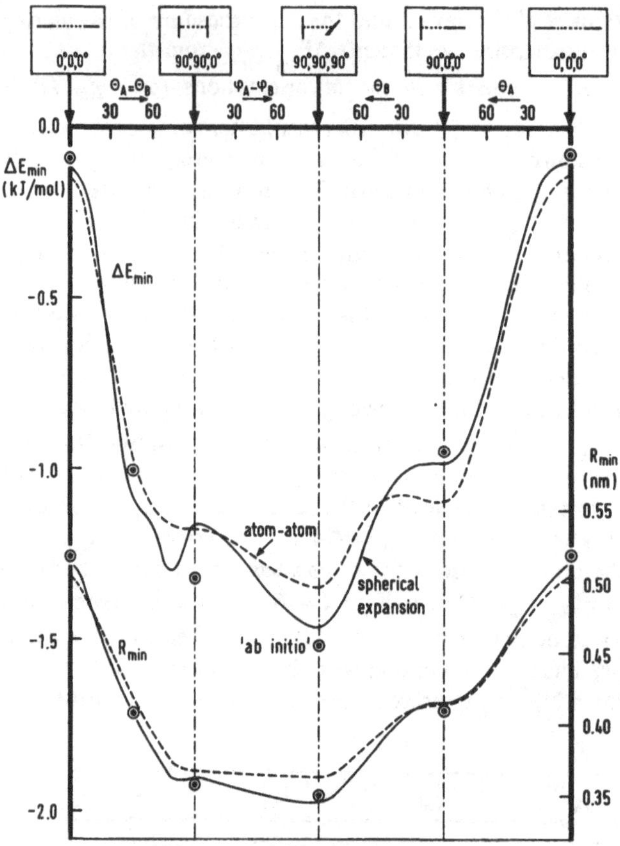

Fig. 10. Orientational dependence of the Van der Waals minimum in the N_2—N_2 interaction potential. The well depth $\Delta E_{min.}$ and equilibrium distance $R_{min.}$ were obtained by varying **R** for each orientation θ_A, θ_B, $\phi_A(\phi_B = 0)$. "Ab initio" results, spherical expansion (4) and atom-atom fit (8) of these results from ref.[136]

Another approach, which was taken by Berns et al.[136] was to fit the "ab initio" interaction energies (for 6 distances and 6 orientations) by an atom-atom potential. This was again done term by term, just as for $(C_2H_4)_2$. The electrostatic (multipole) energy (i) was fitted by a 3 parameter axial point charge model; average error 6.5%. The dispersion (multipole) energy (ii) was fitted by a 2 parameter r^{-6} potential; average error 6.3% (including a slight shift of the origin away from the nuclei; without this shift the error was 9.7%). The overlap (penetration and exchange) energy (iii) was represented by an exponential atom-atom repulsion with 2 parameters (the optimum origins lie practically on the nuclei, in this case); average error 9.2%. The resulting atom-atom potential[136] appears to describe the angular dependence of the N_2—N_2 "ab initio" potential surprisingly well (see figs. 5 and 10), also for the 105 orientations calculated (at R = 3 Å). It is striking that the atom-atom model works much better for N_2—N_2 than for C_2H_4—C_2H_4, especially for the overlap energy. Maybe this is due to the lone-pair electrons in N_2 balancing the effects of chemical bonding. The properties of the N_2 crystal α and γ phases[156,157] (equilibrium structure, cohesion energy, phonon frequencies) are very well described[136,175] by this atom-atom potential.

4.2 Potential Surfaces of Van der Waals Molecules; $(N_2)_2$ and $(C_2H_4)_2$

We have studied the potential hypersurfaces for the Van der Waals dimers $(N_2)_2$ and $(C_2H_4)_2$ by varying all the independent internal coordinates (in the rigid molecule approximation, 3 angles for $(N_2)_2$, 5 angles for $(C_2H_4)_2$ and the distance R in both cases). It is of course not possible to present the complete surfaces pictorially; we have displayed in figs. 9 and 10 some typical cuts through the surface of $(N_2)_2$. Expecially fig. 10 contains much information since the distance was varied to find the energy minimum $\Delta E_{min.}$ for each orientation (θ_A, θ_B, ϕ_A). In the figs. 4 and 5 and 6, 7 and 8 the different contributions ("long range" (i) and (ii) and "short range" (iii), see the previous paragraph) to the interaction energy are plotted. It is not possible to indicate very pronounced minima in the surfaces. For $(N_2)_2$ we have found one absolute minimum at $\Delta E_{min.} = 1.5$ kJ/mol in the crossed structure $\theta_A = \theta_B = \phi_A = 90°$, R = 3.5 Å; for $(C_2H_4)_2$ we have found two equally deep minima at $\Delta E_{min.} = 5.0$ kJ/mol, one for a staggered parallel structure with R = 3.93 Å the other for a skew structure with R = 3.87 Å (see table 4). The potential surfaces are rather flat around these minima, however, and many other possible structures exist with a binding energy higher than 80% of $\Delta E_{min.}$. Therefore, we expect the N_2 and C_2H_4 molecules in the dimers to make rather wide angular oscillations (librations) in some directions. For instance, the barrier for a complete rotation over ϕ_A in the N_2 dimer with $\theta_A = \theta_B = 90°$ is about 0.2 kJ/mol (13% of $\Delta E_{min.}$) with practically no variation of the equilibrium distance (see fig. 10) (the rotational constant of free N_2 is $2.0\ cm^{-1} = 0.024$ kJ/mol). At somewhat higher temperatures where the dimer is still stable, (in some cases maybe at zero temperatures already) these oscillations will go over into (weakly) hindered rotations. In other directions, rotations of the molecules are strongly hindered; the dimer must almost dissociate before such a rotation becomes possible. (e.g. the rotations through the orientation $\theta_A = \theta_B = \phi_A = 0°$ in the N_2 dimer, fig. 10). The solution of the dynamical problem for the nuclei may be not an easy job. Neither the model of almost free rotors, which works very well for the $(H_2)_2$ case[154], nor the harmonic (or weakly anharmonic) oscillator model, which works reasonably well in molecular crystals, may be applicable. In this sense, the problem is comparable to that of the plastic phases in molecular crystals, such as the β-phase of solid N_2.

Leaving aside this dynamical problem, we can make some further remarks about the equilibrium structure of Van der Waals molecules. Some attempts have been made to predict this structure from the molecular properties, multipole moments, polarizabilities, which are reflected in the long range interactions (electrostatic, dispersion). Other authors[161,176] have assumed that the equilibrium structure of Van der Waals dimers resembles the structure of nearest neighbour pairs in molecular crystals. The latter approach could possibly be justified by packing considerations (short range repulsion). An example of the first approach is the prediction of a T-shaped ($\theta_A = 90°$, $\theta_B = \phi_A = 0°$) equilibrium structure for the N_2-dimer, mainly

Table 4a. Most stable dimer structures: N_2-N_2

	R (Å)	θ_A[a]	θ_B[a]	ϕ_A[a]	ΔE (kJ/mol)
dimer[b]	3.64	90°	90°	90°	1.35
$\alpha-N_2$ crystal[c] neighbour pair	3.99 (4.04[d])	90°	35°	55°	1.05
$\gamma-N_2$ crystal[c] neighbour pair	3.79 (3.98[d])	90°	42°	90°	0.94

[a] Angular coordinate system with $\Theta = \Phi = \phi_B = 0°$, see table 1

[b] Minimum in the potential energy surface, neglecting the effects of nuclear motion. Full minimum search was done with the atom—atom potential, fitted to the "ab initio" results[136)]

[c] Experimental crystal structures, see ref.[156)], ΔE calculated with the atom—atom potential[136)]

[d] $R_{min.}$ obtained with the atom—atom potential[136)] for fixed (crystal) orientations

on the basis of attractive quadrupole-quadrupole interactions[158)]. Addition of the anisotropic dispersion interactions[101)] and, even, of the short range repulsion in an approximate model[177)] gives further support for the stability of this T-shaped N_2-N_2 structure, but this also suggests another possible structure of equal stability, the staggered parallel one ($\theta_A = \theta_B \simeq 45°$, $\phi_A = 0°$). A related approximate model which includes the short range repulsions[178)] leads to different predictions, however. Now that we have calculated both the (anisotropic) long range and short range contributions more quantitatively, we can make some more definite, although not final, statements on this matter.

In fig. 5 we see that, indeed, the T-shaped and the staggered parallel structure have maximum electrostatic attraction. The dispersion energy is most favourable, of course, for the linear structure. For distances in the neighbourhood of the (isotropic) Van der Waals minimum the (short range) exchange repulsion is the dominant anisotropic term, however. Since it increases very steeply when the molecular charge clouds start to overlap (especially in the linear structure $\theta_A = \theta_B = \phi_A = 0°$), it determines to a large extent the distance of closest approach of the molecules. If, for a given orientation the long range interactions are not maximally attractive (when compared with other orientations, for equal distance R), but the molecules can approach each other closely, the Van der Waals well may still be relatively deep. This is, for instance, what happens for the crossed structure ($\theta_A = \theta_B = \phi_A = 90°$) of the N_2-dimer. In general, one can observe this role of the short range repulsion from fig. 10, where the well depth $\Delta E_{min.}$ shows a strong (negative) correlation with the equilibrium distance $R_{min.}$. The same phenomenon has been found for the C_2H_4-dimer: maximum binding energy (table 4) occurs for the dimer structures with the smallest $R_{min.}$. Only when the short range repulsion is not very sensitive to a change of orientation (for instance, the rotation over ϕ_A in the N_2-dimer with $\theta_A = \theta_B = 90°$, see fig. 5), the long range interactions (in this case, the electrostatic interactions, even though they are repulsive) can still be important in determining the equilibrium structure.

This crucial role of the short range repulsions (closest packing) for the dimer structure may suggest that the structure of nearest neighbour pairs in the molecular

Table 4b. Most stable dimer structures; $C_2H_4-C_2H_4$

	R (Å)	Θ^a	Φ^a	$\alpha_B{}^a$	$\beta_B{}^a$	$\gamma_B{}^a$	ΔE (kJ/mol)
dimer[b]	3.93	49°	27°	0°	0°	0°	5.01
dimer[b]	3.87	56°	14°	−59°	45°	−59°	5.05
C_2H_4 crystal[c] 1st neighbour pair	4.07 (4.07[d])	61°	13°	0°	0°	0°	3.89
C_2H_4 crystal[c] 2nd neighbour pair	4.44 (4.50[d])	54°	7°	−47°	96°	47°	2.52

[a] Angular coordinate system with $\alpha_A = \beta_A = \gamma_A = 0°$; α_B, β_B, γ_B Euler angles as defined in ref.[18]; molecular axes: C—C = y-axis, C_2H_4-plane = xy-plane

[b] Lowest minima in the atom—atom potential energy surface, fitted to the ab initio results[155]

[c] Experimental crystal structure from ref.[170], ΔE calculated with the atom—atom potential[155]

[d] $R_{min.}$ obtained with the atom—atom potential[155] for the fixed (crystal) orientations.

crystal indeed forms a good indication for the structure of the Van der Waals dimer. This is supported by one of the stablest C_2H_4 dimers, the one with the staggered parallel structure, which resembles rather closely one of the neighbour pairs in the crystal. The other most stable C_2H_4 dimer and, also, the crossed N_2 dimer deviate more strongly from the crystal neighbour configurations, however (see table 4). And, in fact, it is not obvious, even if only packing considerations determine the structure, that the optimum packing in a crystal where each molecule is surrounded by several neighbours must correspond with optimally packed dimers. The crystal neighbours should not have too unfavorable pair energies, though, and we have checked on our $(C_2H_4)_2$ and $(N_2)_2$ potential surfaces that this is not the case (see table 4).

Summarizing this section on the potential surfaces of $(N_2)_2$ and $(C_2H_4)_2$ we try to make a few concluding remarks which may be more generally applicable. Clearly, this generalization, if valid at all, is restricted to Van der Waals complexes composed of molecules which have zero or small dipole moments (excluding, for instance, hydrogen bonded systems), where the dispersion energy is the dominant cohesive contribution. Sometimes, these complexes have been called Van der Waals molecules in the proper sense. We expect the N_2 and C_2H_4 dimers to be typical examples of such complexes. The equilibrium structure is, in first instance, determined by minimal short range repulsions (packing, steric hindrance considerations). If these allow several rather closely packed structures with little differences in energy, then the long range anisotropic interactions (both electrostatic and dispersion) will determine the lowest energy configuration. The balance between the different attractive and repulsive contributions can be very subtle, however, and there may be several competing dimer configurations with almost equal binding energies. (Such as we have found both for $(N_2)_2$ and for $(C_2H_4)_2$.) It becomes very hard then to predict the stablest structure, even on the basis of ab initio calculations. (More approximate model calculations are completely useless in this respect). Moreover, one has to take into account the (nuclear) dynamical problem. Only in combination with experimental information the calculations may provide conclusive answers about the structure of Van der Waals complexes. For instance, the technique of molecular beam deflection[159] is sensitive to the dipole moment of a Van der Waals molecule (cf. sect. 5). One of the calculated stablest structures of the C_2H_4 dimer, the staggered parallel one, has a vanishing dipole moment because of symmetry. For the skew structure this is not the case. So, the experiment, if it is sufficiently sensitive, could discriminate between these structures. For the N_2 dimer the stablest structure we have calculated, the crossed one, has zero dipole moment, but so has the staggered parallel one, which we find only slightly higher in energy. Here, the absence of a dipole moment is not sufficient to decide which structure is more favourable and additional experimental information is needed to check our result. On the other hand, the experimental data alone are mostly not sufficient to obtain a detailed potential surface for Van der Waals molecules; the ab initio results, wherever they can be obtained, are very helpful for interpreting the experimental observations.

5 Some Properties of Van der Waals Molecules

5.1 Orientational Dependence

In the very same way as the Born-Oppenheimer approximation allows the definition of a potential energy surface for a Van der Waals molecule, it enables, too, the concept of an interaction tensor field. This is a field dependent on the relative coordinates of the monomers and transforming as a tensor under rotation of the complex as a whole. (The potential energy surface is an example of a rank zero interaction tensor field). In the case of tensor fields it is also convenient to base the theory on irreducible tensors and to use an expansion in terms of a complete set of functions of the five angular coordinates describing a Van der Waals dimer.

The generalization of the scalar-valued angular functions (1b) to arbitrary rank J is:

$$A^J_{\Lambda,N}(\underline{\omega}_A, \underline{\omega}_B, \vec{R}) = \sum_{M_A, M_B, M} (L_A, M_A; L_B, M_B; L, M \mid J, N) \times D^{L_A}_{M_A, K_A}(\underline{\omega}_A)^* D^{L_B}_{M_B, K_B}(\underline{\omega}_B)^* C^L_M(\underline{\Omega}) \tag{43}$$

where the generalized Clebsch-Gordan coefficient is given by

$$(L_A, M_A; L_B, M_B; L, M \mid J, N) = \sum_{\mu=-\lambda}^{+\lambda} (L_A, M_A; L_B, M_B \mid \lambda, \mu)\, (\lambda, \mu; L, M \mid J, N)\,. \tag{44}$$

The set $\{A^J_{\Lambda,N} | N = -J, \ldots, +J\}$ transforms cogrediently to the set of spherical harmonics of order J. Using

$$(L_A M_A; L_B M_B; LM \mid 00) = \begin{pmatrix} L_A & L_B & L \\ M_A & M_B & M \end{pmatrix} (-1)^{L_A - L_B + L} \tag{45}$$

we indeed find the expression (1b) as a special case of (43), apart from a sign.

An irreducible interaction tensor Θ^J of order J can be expanded as:

$$\Theta^J_N(\underline{\omega}_A, \underline{\omega}_B, \vec{R}) = (2J+1)^{-1/2} \sum_{\Lambda} T^J_\Lambda(R)\, A^J_{\Lambda,N}(\underline{\omega}_A, \underline{\omega}_B, \underline{\Omega})\,, \tag{46}$$

where Λ is the set of quantum numbers defined in table 1. Because of the Wigner-Eckart theorem the expansion coefficients $T^J_\Lambda(R)$ (reduced matrix elements) are independent of the magnetic quantum number N.

It is possible to apply the multipole expansion and perturbation theory in order to derive long range expressions for $T^J_\Lambda(R)$, thus relating this quantity to monomer properties. A simple example of such a procedure can be found in the appendix of ref.[179], where the induction contribution to the dipole moment ($J = 1$) of an arbitrary Van der Waals dimer has been evaluated.

Because not much is known experimentally about general interaction tensors, and especially not about their long range behaviour, we will not pursue this line of approach, but rather give a brief review of the existing work which has concentrated on two different tensors: the pair dipole (order 1 tensor) and the pair polarizability (order 2 plus order 0 tensor).

5.2 Interaction Dipole Moments

The dipole moment of a Van der Waals dimer consists in principle of three contributions: the dipole moments of the two monomers and the interaction dipole moment. In the usual Van der Waals molecules the interaction dipole is in the order of 0.1 D[159] which for a large part arises from induction. That is, permanent moments on the one monomer induce a dipole moment on the other. Obviously, this effect is absent in the case of a dimer consisting of two noble gases. Here the interaction dipole moment is an order of magnitude smaller and is largely due to the short range effects exchange and penetration[180].

The measurement of interaction dipoles by beam deflection[181] gives an indication of the structure of the Van der Waals molecule. A recent example is given by Howard and coworkers[182], who experimentally established that the interaction dipole of $(CO_2)_2$ is less than 10^{-2} D. Since it has sometimes been suggested[159,183] that the dimer is a T-shaped complex (a favourable configuration for quadrupole-quadrupole interaction, see above), they estimated the induction contribution to the dipole moment for this conformation (at $R = 4.1$ Å) and found 0.18 D for this value. So they conclude that the dimer has most likely a staggered parallel configuration, which is in accordance with recent ab initio calculations[184].

Another experimental source for interaction dipoles is the measurement of pressure induced absorption[185]. Strictly speaking this effect does not belong to the realm of Van der Waals complexes, because one measures here infrared radiation absorbed by *unbound* complexes. But since much can be learned about Van der Waals interactions from an interpretation of the data, we briefly review the work in this area.

First it should be noted that most of the experimental work on infrared absorption of gas mixtures has been restricted to cases where the constituent molecules themselves are not infrared active. Much work has for instance been done on mixtures of noble gases[186] and noble gases with H_2[187]. Noble gas mixtures show a broad band centered around 100 cm^{-1}. This is due to absorption by the translational motion of two unlike atoms relative to their joint center of mass. The same kind of translational band has also been measured in H_2-noble gas mixtures[187,188] and pure H_2 [188]

Also rotational bands, with transitions lying in the region from 370 to 810 cm^{-1}, have been observed[188]. It is parenthetically interesting to note that the study of the translational band in $(H_2)_2$ and H_2—He is of astrophysical interest, as the greenhouse effect on the outer planets is believed to be largely due to the translational band of these two dimers[189]. (The temperature at the surface of these planets is about 150 °K $\simeq$ 100 cm^{-1}). In the case of gases containing hydrogen one also observes the fundamental $v = 0 \rightarrow 1$ band. This transition becomes (weakly) allowed under the influence of the interaction with the collision partner.

Much effort has been put into the explanation of the spectral line shapes[190], but it seems that the definite theory has yet to be established. In the meantime one can extract useful information from the first few moments of the spectral density, by applying the elegant theory developed by Van Kranendonk[191] and Poll and Van Kranendonk[192]. This theory relates the first moment to the derivative of the dimer dipole moment with respect to the intermolecular distance. The zeroth moment yields information about the square of the dipole moment. As this review is not the place to go extensively into the Van Kranendonk theory, we only note that, once the intermolecular potential surface and the interaction dipole field are known, — for instance by ab initio calculations — it is relatively easy to compute the moments of the spectral density. Since these are directly observable, the experiment of pressure induced absorption may serve as a check on the correctness of ab initio calculations, not only of the interaction energy, but also of the interaction dipole.

The first ab initio calculations on interaction dipoles were performed by Matcha and Nesbet[193]. They considered the systems HeNe, HeAr and NeAr as "supermolecules" and did ordinary Hartree-Fock-LCAO-SCF calculations in the range R = 2.0 to 5.5a_0. Because of the Hartree-Fock approximation they did not obtain the dispersion contribution to the dipole moment (cf. sect. 2), but only exchange, penetration and overlap-induction contributions. Their ab initio dipoles could be fitted quite well by a single exponential, which supported the assumption made earlier by Van Kranendonk[191].

Later Byers Brown and Whisnant considered in detail the importance of dispersion, first theoretically[194] by deriving Unsöld type expressions for the leading R^{-7} terms and subsequently numerically for HeH and HeHe. (In the case of HeHe the atomic contributions add up to zero, of course). At around the same time Lacey and Byers Brown[195] considered also exchange and penetration contributions (in first order of perturbation theory) in addition to dispersion. They considered HeNe, HeAr, NeAr and ArKr in the range 4.0–9.0a_0. Since the collision induced absorption is largely due to complexes with intermolecular distances close to the scattering diameter σ[196], it is interesting to compare the values of the exchange dipole and the dispersion dipole at $R = \sigma$, (although the long range approximations which lead to the dispersion values are subject to serious doubt at such a short distance). Lacey and Byers Brown find that, except for HeNe the two contributions have opposite sign and that the dispersion dipole is about an order of magnitude smaller than that due to exchange.

Recent calculations by Berns et al.[179] show that also for the He—H_2 system the dispersion contribution is small. At long range it is completely dominated by induction, at short range by overlap effects. The calculations of Berns et al. have been performed by the VB approach mentioned above[63]. No perturbation theory or multipole expansion was applied, although the VB method makes it possible to interpret the results in the usual long range terms plus exchange and penetration. In this manner it was found that the VB contribution which corresponds in the long range to the R^{-7} dispersion term completely failed to have an R^{-7} dependence from 3.7 Å ($7a_0$) inward. This was surprising as the corresponding VB induction term kept its R^{-4} behaviour, predicted by long range theory, to much shorter distances. In any case, these VB calculations have shown that a very good description of the interaction induced dipole of the HeH_2 complex requires only the inclusion of first order exchange, charge cloud penetration and — as the only second order term — induction. Since all these terms are accounted for in the supermolecule SCF approach such an approach seems to be ideally suited for routinely obtaining accurate dipole moments. However, when such calculations were undertaken in our institute[196] a (somewhat unexpected) difficulty arose. Interaction dipole moments appeared to be much more sensitive to the basis set superposition error than interaction energies, and hence large and well balanced basis sets had to be employed. It may well be that inadequate basis sets form the source for the unreliability of the Matcha-Nesbet results[193] at larger R-values.

However, once one is aware of the problem the basis set superposition error can easily be checked by a ghost molecule treatment[63, 108–110]. Proceeding in this manner it was found[196] that the spectral moments computed from the SCF results for HeH_2 led to good agreement with the available experimental data. The outcome of the ab initio calculations also suggested parameters in the analytic representation of the dipole moment, which in a few respects differed considerably from those used so far in the interpretation of the experiments. When these new parameteres will indeed prove to describe the experiment better than the existing ones, it will be yet another example of how the interplay of ab initio calculations and experimental work can be useful.

5.3. Pair polarizabilities

The influence of Van der Waals interactions on the polarizability of interacting molecules manifests itself in deviations from the Clausius-Mosotti equation[197], in the Kerr effect[198] and in collision induced light scattering[199]. Although measurements of these effects are all performed on bulk systems in thermodynamical equilibrium and not on Van der Waals molecules per se, we will nevertheless say a few words about pair polarizabilities, because, just as in the case of the collision induced IR absorption, much can be learned about Van der Waals interactions from the comparison of experimental and computational results.

In a pioneering paper[200] Jansen and Mazur established the quantum mechanical basis for the effect of molecular interactions on the polarizability of spherical atoms. Using long range theory (no intermolecular exchange, Rayleigh-Schrödinger perturbation theory and only the first term in a multipole expansion of the intermolecular interaction) they derived an expansion of the pair polarizability as a power series in R^{-1}. The first two terms (in R^0 and R^{-3}) are the same as those obtained from classical electrostatics, the quantum mechanical effect of dispersion appears in the third (R^{-6}) and higher terms. In a subsequent paper[197] Mazur and Jansen applied their result to the dielectric constant appearing in the Clausius-Mosotti equation, making a virial (density) expansion of this constant. They showed that the dielectric constant depends on the trace of the pair polarizability tensor, or rather, on the change in this trace caused by varying R. So, the dielectric constant provides information about the isotropic (J = 0) part of the polarizability tensor. (It is interesting to note that the depolarized Raman intensity depends on the anisotropic (J = 2) part of the polarizability tensor, and hence it is also experimentally convenient to separate the polarizability tensor into irreducible components.) When later dielectric second virial coefficients B_ε were measured[201,202] for He, Ne, Ar, Kr, H_2 and N_2, it was found that the results predicted by long range theory were considerably at variance with the experimental findings. It was therefore suggested[201] that short range effects could not at all be neglected. And indeed, a finite field Hartree-Fock supermolecule calculation on He_2, (including exchange and penetration)[203] yields $B_\varepsilon = -0.093\ \mathrm{cm^6\ mol^{-2}}$ at room temperature, compared to the experimental result $B_\varepsilon = -0.06 \pm 0.04$. (The long range result has a positive sign, indicating that long range theory predicts the isotropic polarizability α to increase with decreasing distance R, whereas α decreases[203] in the range of physical interest.)

Other ab initio calculations on the Hartree-Fock level of B_ε for He_2 [204-206] also gave good agreement with the experimental results at room temperature, but at 4 °K a serious disagreement between theory and experiment appeared, which cannot be explained by approximations in the calculations. It has been suggested that the experimental data at 4 °K have to be reinterpreted[204]. Recent ab initio calculations including correlation[207] give $B_\varepsilon = -0.06\ \mathrm{cm^6\ mol^{-2}}$ at 322 °K; so this value is now firmly established, experimentally as well as theoretically.

Whereas the dielectric constant probes essentially the R-dependence of the *isotropic* polarizability, the collison induced depolarized Raman scattering depends on the increment in the *anisotropy* of the polarizability with varying R. Depolarized Raman scattering of noble gases has first been observed by MacTague and Birnbaum[208] in 1968, and later investigated for many gases[209]. Very recently also polarized Raman intensities have been measured for Ne_2 [210,211] and He_2 [212,213].

It is remarkable that short range forces, such as exchange and penetration, seem to have much less influence on the anisotropic than on the isotropic part of the polarizability. This has been observed in the interpretation of Raman data[213] as well as in the results of ab initio calculations including correlation[207].

Several calculations of the polarizability tensor of noble gas dimers have been made[214-216] which do include charge penetration, but not exchange. The work by Oxtoby and Gelbart[214] is based on the concept of polarizability density. However, as pointed out by Sipe and Van Kranendonk[217], this concept, borrowed from macroscopic dielectric theory may lead to erroneous results for moments of order higher than 1. Similar criticism has been raised by Buckingham and coworkers[216], who have introduced instead a model based on perturbed (by the external field) atomic charge densities. They have calculated the collision induced polarizabilities of He_2 and Ar_2, without exchange, and have found an anisotropy which is in excellent agreement with recent experimental data for He_2 [213] exhibiting again that exchange does not affect the anisotropy much at distances of physical interest.

Finally, it must be pointed out that theory and experiment are not yet in complete agreement with regard to the trace of the pair polarizability of He_2. The most complete quantum chemical treatment to date[207], one including correlation, is not fully consistent with recent polarized Raman data[213], which is surprising, since less complete (Hartree-Fock level) quantum chemical treatments[203, 204, 218] account very well for the observed data[213]. This is the more surprising as the correlation calculations give complete agreement with the experimentally determined second dielectric virial coefficient B_ε, whereas the calculations on the Hartree-Fock level are here off by about 30% (see above). An explanation for this discrepancy can perhaps be found in the fact that B_ε depends linearly on the trace of the polarizability tensor, whereas the polarized Raman intensities are proportional to the square of this trace. Hence the two experiments constitute different tests on the trace. In the case of Ne_2 there is still considerable disagreement between experiment[211] and ab initio calculations[218], for the isotropic as well as for the anisotropic parts of the polarizability.

Acknowledgement We thank dr. Tadeusz Luty for many stimulating discussions.

6 Appendix

Proof of the invariance of the function $A_\Lambda(\underset{=}{\omega}_A, \underset{=}{\omega}_B, \underset{=}{\Omega})$

Consider a molecule with orientation $\underset{=}{\omega}_1 \equiv \{\alpha_1, \beta_1, \gamma_1\}$. When we rotate this molecule over the Euler angles $\underset{=}{\omega}$, the set of Euler angles $\underset{=}{\omega}_2$ describing the new orientation of the molecule, may be obtained from the matrix equation:

$$\underset{\equiv}{R}(\underset{=}{\omega}_2) = \underset{\equiv}{R}(\underset{=}{\omega})\, \underset{\equiv}{R}(\underset{=}{\omega}_1)\,, \tag{A1}$$

where $\underset{\equiv}{R}(.) \in SO(3)$ stands for a 3×3 rotation matrix.

The Wigner D-matrices, defined in equation (2), belong to the Hilbert space $L^2[SO(3)]$. With a rotation $\underset{=}{\omega}$ of the molecule one can associate an operator $\hat{R}(\underset{=}{\omega})$ on this Hilbert space by defining:

$$\hat{R}(\underset{=}{\omega})\, \underset{\equiv}{D}^L(\underset{=}{\omega}_2) \equiv \underset{\equiv}{D}^L(\underset{=}{\omega}_1) \tag{A2}$$

(This is Wigner's convention.) Realizing that $\underset{\equiv}{D}^L(.)$ is a short-hand notation for $\underset{\equiv}{D}^L(\underset{\equiv}{R}(.))$, and recalling that $\underset{\equiv}{D}$ is a representation of SO(3), we find, invoking (A1),

$$\hat{R}(\underset{=}{\omega})\, \underset{\equiv}{D}^L(\underset{=}{\omega}_2) = \underset{\equiv}{D}^L(\underset{=}{\omega})^{-1} \underset{\equiv}{D}^L(\underset{=}{\omega}_2) \tag{A3}$$

or:

$$\hat{R}(\underset{=}{\omega})\, D^L_{M,K}(\underset{=}{\omega}_2) = \sum_{M'} D^L_{M',M}(\underset{=}{\omega})^*\, D^L_{M',K}(\underset{=}{\omega}_2)\,. \tag{A4}$$

From (A4) we draw the important conclusion that every column of a D-matrix is an irreducible tensorial set of order L, transforming contragrediently to the set of spherical harmonics of the same order. Indeed, if we take K = 0, and use that[17]:

$$D^{L}_{M,0}(\alpha, \beta, \gamma) = C^{L}_{M}(\beta\alpha)^{*}, \tag{A5}$$

we find the complex conjugate of the usual transformation equation for spherical harmonics.

The Wigner 3j-symbol is often defined as the coefficient coupling a product of three irreducible tensors (of the same variance) to an invariant[46]. Invoking this definition, it immediately follows that the function $A_{\Lambda}(\underline{\omega}_A, \underline{\omega}_B, \underline{\Omega})$ is an invariant.

However, a more explicit proof is obtained by rotating the D-matrices and the spherical harmonics appearing in the definition (1b) of $A_{\Lambda}(\underline{\omega}_A, \underline{\omega}_B, \underline{\Omega})$ by using Eq. A4, and subsequent application of the following relation[17]:

$$\begin{pmatrix} L_A & L_B & L \\ M'_A & M'_B & M' \end{pmatrix} = \sum_{M_A, M_B, M} D^{L_A}_{M'_A, M_A}(\underline{\omega})\, D^{L_B}_{M'_B, M_B}(\underline{\omega})\, D^{L}_{M', M}(\omega) \begin{pmatrix} L_A & L_B & L \\ M_A & M_B & M \end{pmatrix} \tag{A6}$$

This shows that rotation of the dimer over $\underline{\omega}$ leaves the function $A_{\Lambda}(\underline{\omega}_A, \underline{\omega}_B, \underline{\Omega})$ invariant.

7 References

1. Ewing, G. E.: Can. J. Phys. *54*, 487 (1976); Howard, B. J., in: Physical Chem. Ser. 2, Vol. 2, Molecular Structure and Properties, MTP Internat. Rev. Sci., Butterworths, London (1975); LeRoy, R. J., Carley, J. S.: Adv. Chem. Phys., 1980
2. Hirschfelder, J. O., Curtiss, C. F., Bird, R. B.: Molecular Theory of gases and liquids, Wiley, New York (1974)
3. Intermolecular Forces, Discuss. Faraday Soc. *40* (1965)
4. Hirschfelder, J. O. (ed): Adv. Chem. Phys. *12* (1967)
5. Margenau, H., Kestner, N. R.: Theory of Intermolecular Forces, 2 ed., Pergamon, New York (1971)
6. Certain, P. R., Bruch, L. W., in: Physical Chem. Ser. *1*, Vol. 1, Theoretical Chemistry, MTP Internat. Rev. Sci., Butterworths, London (1972)
7. Amos, A. T., Crispin, R. J., in: Theoretical Chemistry, Advances and Perspectives, Vol. 2 (H. Eyring, D. Henderson eds.), Academic Press, New York (1976)
8. Murrell, J. N., in: Rare Gas Solids (M. L. Klein, J. A. Venables eds.), Academic Press, London (1976)
9. Intermolecular Interactions: From Diatomics to Biopolymers (B. Pullman ed.), Wiley, New York (1978)
10. Kihara, T.: Intermolecular Forces, Wiley, New York (1978)
11. Hobza, P., Zahradnik, R.: Weak Intermolecular Interactions in Chemistry and Biology, Elsevier, Amsterdam (1980)
12. Atom-Molecule Collision Theory (R. B. Bernstein ed.), Plenum, New York (1979)
13. Kaplan, I. G.: Intern. J. Quantum Chem. 16, 445 (1979)
14. Hirschfelder, J. O., Meath, W. J.: ref. 4, p. 3
15. Steele, W. A.: J. Chem. Phys. *39*, 3197 (1963); Stone, A. J.: Mol. Phys. *36*, 241 (1978); Stone, A. J., in: Molecular Physics of Liquid Crystals, (G. R. Luckhurst, G. W. Gray eds.), Academic Press, New York (1979); Egelstaff, P. A., Gray, C. G., Gubbins, K. E., in: Physical Chem. Ser. 2, Vol. 2, Molecular Structure and Properties, MTP Internat. Rev. Sci., Butterworths, London (1975)

16. Stolte, S., Reuss, J., in ref. 12, p. 201 (Note that these authors use a different definition of the D-matrices (2)). Secrest, D., in ref. 12, p. 265, (The expansion formula in this paper is slightly incorrect)
17. Brink, D. M., Satchler, G. R.: Angular Momentum, Clarendon Press, Oxford, 2 ed., (1975)
18. Messiah, A.: Quantum Mechanics, Appendix C. North Holland Amsterdam, (1965)
19. Barut, A. O., Raczka, R.: Theory of Group Representations and Applications, PWN. — Polish Scientific Publishers, Warszawa (1977)
20. Mulder, F., van der Avoird, A., Wormer, P. E. S.: Mol. Phys. *37*, 159 (1979)
21. Pauly, H., in ref. 12, p. 111
22. LeRoy, R. J., van Kranendonk, J.: J. Chem. Phys. *61*, 4570 (1974)
23. LeRoy, R. J., Carley, J. S., Grabenstetter, J. E.: Faraday Discuss. Chem. Soc. *62*, 169 (1977); Carley, J. S.: ibid. *62*, 303 (1977); Carley, J. S.: Thesis, Waterloo (1978)
24. Dyke, T., Howard, B. J., Klemperer, W.: J. Chem. Phys. *56*, 2442 (1972)
25. Dunker, A. M., Gordon, R. G.: ibid. *68*, 700 (1978)
26. Zandee, L., Reuss, J.: Chem. Phys. *26*, 327, 345 (1977)
27. Thuis, H., Stolte, S., Reuss, J.: Comments Atom. Mol. Phys. *8*, 123 (1979); Thuis, H.: Thesis, Nijmegen (1979)
28. Buck, U., Khare, V., Kick, M.: Mol. Phys. *35*, 65 (1978)
29. Ahlrichs, R., Penco, R., Scoles, G.: Chem. Phys. *19*, 119 (1977)
30. Tang, K. T., Toennies, J. P.: J. Chem. Phys. *68*, 5501 (1978)
31. Barker, J. A., in: Rare gas Solids (M. L. Klein, J. A. Venables eds.), Academic Press, London (1976)
32. Walmsley, S. H., Pople, J. A.: Mol. Phys. *8*, 345 (1964); Raich, J. C., Gillis, N. S., Anderson, A. B.: J. Chem. Phys. *61*, 1399 (1974)
33. Koide, A., Kihara, T.: Chem. Phys. *5*, 34 (1974)
34. LaBudde, R. A., Bernstein, R. B.: J. Chem. Phys. *55*, 5499 (1971); Keil, M., Parker, G. A., Kuppermann, A.: Chem. Phys. Lett. *59*, 443 (1978)
35. McRury, T. B., Steele, W. A., Berne, B. J.: J. Chem. Phys. *64*, 1288 (1976)
36. Price, S. L., Stone, A. J.: Mol. Phys. *40*, 805 (1980)
37. Downs, J. et al.: Mol. Phys. *37*, 129 (1979)
38. Sack, R. A.: J. Math. Phys. *5*, 260 (1964)
39. Buehler, R. J., Hirschfelder, J. O.: Phys. Rev. *83*, 628 (1951)
40. Rose, M. E.: J. Math. Phys. *37*, 215 (1958)
41. Wormer, P. E. S.: Thesis, Nijmegen (1975)
42. Gray, C. G., Lo, B. W. N.: Chem. Phys. *14*, 73 (1976)
43. Young, R. H.: Intern. J. Quantum Chem. *9*, 47 (1975); Jansen, L.: Phys. Rev. *110*, 661 (1958)
44. Ahlrichs, R.: Theoret. Chim. Acta *41*, 7 (1976)
45. Wormer, P. E. S., Mulder, F., van der Avoird, A.: Intern. J. Quantum Chem. *11*, 959 (1977)
46. Fano, U., Racah, G.: Irreducible Tensorial Sets, Academic Press, New York (1959)
47. Chalasinsky, G., Jeziorski, B., Szalewicz, K.: Intern. J. Quantum Chem. *11*, 247 (1977)
48. Claverie, P.: ibid *5*, 273 (1971)
49. Johnston, D. F.: Repts. Progr. Phys. *23*, 66 (1960)
50. Matsen, F. A.: Adv. Quantum Chem. *1*, 59 (1964)
51. Wormer, P. E. S., van der Avoird, A.: J. Chem. Phys. *57*, 2498 (1972); Intern. J. Quantum Chem. *8*, 715 (1974)
52. Eisenschitz, R., London, F.: Z. Phys. *60*, 491 (1930)
53. Chipman, D. M., Bowman, J. D., Hirschfelder, J. O.: J. Chem. Phys. *59*, 2830 (1973)
54. Jeziorski, B., Kolos, W.: Intern. J. Quantum Chem. *12S1*, 91 (1977)
55. Adams, W. H., Polymeropoulos, E. E.: Phys. Rev. A, *17*, 11, 18, 24 (1978)
56. Kutzelnigg, W.: Intern. J. Quantum Chem. *14*, 101 (1978)
57. Chipman, D. M., Hirschfelder, J. O.: J. Chem. Phys. *59*, 2838 (1973)
58. Jeziorski, B., Szalewicz, K., Chalasinski, G.: Intern. J. Quantum Chem. *14*, 271 (1978)
59. Chalasinski, G., Jeziorski, B.: Theoret. Chim. Acta, *46*, 277 (1977)
60. Schaefer, H. F. (ed.): Methods of Electronic Structure Theory, Plenum, New York (1977); Applications of Electronic Structure Theory, Plenum, New York (1977)

61. Van Duijneveldt-Van de Rijdt, J. G. C. M., Van Duijneveldt, F. B.: Chem. Phys. Lett. *17*, 425 (1972)
62. Margenau, H., Murphy, G. M.: The Mathematics of Physics and Chemistry, 2 ed., Van Nostrand, Princeton (1956)
63. Wormer, P. E. S., van der Avoird, A.: J. Chem. Phys. *62*, 3326 (1975)
64. Kitaura, K., Morokuma, K.: Intern. J. Quantum Chem. *10*, 325 (1976)
65. Chalasinski, G., Jeziorski, B.: Mol. Phys. *27*, 649 (1974)
66. Kestner, N. R.: J. Chem. Phys. *48*, 252 (1968)
67. Wormer, P. E. S., van Berkel, T., van der Avoird, A.: Mol. Phys. *29*, 1181 (1975)
68. Daudey, J. P., Malrieu, J. P., Rojas, O.: Intern. J. Quantum Chem. *8*, 17 (1974)
69. Mulder, F., Geurts, P. J. M., van der Avoird, A.: Chem. Phys. Lett. *33*, 215 (1975)
70. Murrell, J. N., Shaw, G.: J. Chem. Phys. *49*, 4731 (1968)
71. Daudey, J. P., Novaro, O., Berrondo, M.: Chem. Phys. Lett. *62*, 26 (1979) and references therein; Bulski, M., Chalasinski, G.: to be published
72. Claverie, P., in ref. 9, p. 69
73. Kutzelnigg, W.: Faraday Disc. Chem. Soc. *62*, 185 (1977)
74. Werner, H. J., Meyer, W.: Mol. Phys. *31*, 855 (1976)
75. Mulder, F. et al.: Theoret. Chim. Acta, *46*, 39 (1977)
77. Bartlett, R. J., Purvis, G. D.: Physica Scripta, Topical Issue on Many-Body Theory of Atomic Systems (1980)
78. Mulder, F., Berns, R. M.: Internal Report Inst. Theoret. Chem., Nijmegen (1978)
79. Dalgarno, A.: Adv. Chem. Phys. *12*, 143 (1967)
80. Hirschfelder, J. O., Byers Brown, W., Epstein, S. T.: Adv. Quantum Chem. *1*, 255 (1964)
81. Langhoff, P. W., Karplus, M., in: The Padé Approximant in Theoretical Physics (G. A. Baker, J. L. Gammel eds.), Academic Press, New York (1970), p. 41
82. Unsöld, A.: Z. Phys. *43*, 563 (1927)
83. London, F.: Z. Phys. Chem. (B) *11*, 222 (1930)
84. Mulder, F., Huiszoon, C.: Mol. Phys. *34*, 1215 (1977)
85. Mulder, F., van Dijk, G., Huiszoon, C.: Mol. Phys. *38*, 577 (1979); *40*, 247 (1980)
86. Lekkerkerker, H. N. W., Coulon, Ph., Luyckx, R.: Physica *88A*, 375 (1977)
87. Luyckx, R., Coulon, Ph., Lekkerkerker, H. N. W.: J. Chem. Phys. *70*, 4212 (1979); *71*, 3462 (1979); Luyckx, R. et al.: Phys. Rev. *A19*, 324 (1979)
88. Luyckx, R.: Thesis, Brussel (1979)
89. Rivail, J. L., Cartier, A.: Mol. Phys. *36*, 1085 (1978)
90. Hylleraas, E. A.: Z. Phys. *65*, 209 (1930)
91. Kirkwood, J. G.: Physik Z. *33*, 57 (1932)
92. Casimir, H. B. G., Polder, D.: Phys. Rev. *73*, 360 (1948)
93. Langhoff, P. W., Karplus, M.: J. Chem. Phys. *53*, 233 (1970); Langhoff, P. W., Gordon, R. G., Karplus, M.: J. Chem. Phys. *55*, 2126 (1971)
94. Pack, R. T.: J. Chem. Phys. *61*, 2091 (1974); Parker, G. A., Pack, R. T.: J. Chem. Phys. *64*, 2010 (1976)
95. Zeiss, G. D., Meath, W. J.: Mol. Phys. *30*, 161 (1975), *33*, 1155 (1977); Thomas, G. F., Meath, W. J.: Mol. Phys. *34*, 113 (1977)
96. Tang, K. T., Norbeck, J. M., Certain, P. R.: J. Chem. Phys. *64*, 3063 (1976)
97. Nesbet, R. K.: Phys. Rev. *A14*, 1065 (1976)
98. Briggs, M. P., Murrell, J. N., Stamper, J. G.: Mol. Phys. *17*, 381 (1969)
99. Margoliash, D. J., Meath, W. J.: J. Chem. Phys. *68*, 1426 (1978)
100. Dalgarno, A., Morrison, I. H., Pengelly, R. M.: Intern. J. Quantum Chem. *1*, 161 (1967)
101. Mulder, F., van Dijk, G., van der Avoird, A.: Mol. Phys. *39*, 407 (1980)
102. Koide, A.: J. Phys. B: Atom. Molec. Phys. *9*, 3173 (1976)
103. Hoffmann, R.: J. Chem. Phys. *39*, 1397 (1963); Anderson, A. B.: J. Chem. Phys. *64*, 2266 (1976)
104. Pople, J. A., Beveridge, D. L.: Approximate Molecular Orbital Theory, McGraw Hill, New York (1970); Hashimoto, M., Isobe, T.: Bull. Chem. Soc. Japan *46*, 2581 (1973)
105. Gordon, R. G., Kim, Y. S.: J. Chem. Phys. *56*, 3122 (1972); Parker, G. A., Pack, R. T.: ibid. *69*, 3268 (1978); Davis, S. L., Boggs, J. E., Mehrotra, S. C.: ibid. *71*, 1418 (1979)
106. Price, S. L., Stone, A. J.: Chem. Phys. Lett. *65*, 127 (1979)

107. Römelt, J., Peyerimhoff, S. D., Buenker, R. J.: Chem. Phys. *34*, 403 (1978)
108. Boys, S. F., Bernardi, F.: Mol. Phys. *19*, 553 (1970)
109. Groen, T. P., van Duijneveldt, F. B.: Theoret. Chim. Acta, submitted for publication
110. Bulski, M., Chalasinski, G.: ibid., *44*, 399 (1977)
111. Dacre, P.: Mol. Phys. *37*, 1529 (1979); Chem. Phys. Lett. *50*, 147 (1977)
112. Malrieu, J. P., Spiegelman, F.: Theoret. Chim. Acta, *52*, 55 (1979)
113. McWeeny, R., Sutcliffe, B. T.: Methods of Molecular Quantum Mechanics, p. 137, Academic Press, London (1969)
114. Liu, B., McLean, A. D.: unpublished results quoted in refs. 8 and 31
115. Hariharan, P. C., Kutzelnigg, W.: Progr. Rep. Lehrstuhl Theoret. Chemie, Universität Bochum (1977)
116. Schaefer, J., Meyer, W.: J. Chem. Phys. *70*, 344 (1979); Meyer, W.; unpublished results
117. Maeder, F., Kutzelnigg, W.: Chem. Phys. Lett. *37*, 285 (1976)
118. Snyder, L. C., Basch, H.: Molecular Wave Functions and Properties, Wiley, New York (1972)
119. Richards, W. G. et al., Bibliography of Ab Initio Molecular Wave Functions, Oxford University Press, Oxford, (1971); Suppl. for 1970–73, Oxford University Press, Oxford (1974); Suppl. for 1974–77, Clarendon Press, Oxford (1978)
120. Margenau, H.: J. Chem. Phys. *6*, 896 (1938)
121. Fontana, P. R.: Phys. Rev. *123*, 1865 (1961)
122. Davison, W. D.: J. Phys. B. Atom. Molec. Phys. *1*, 139 (1968)
123. Starkschall, G., Gordon, R. G.: J. Chem. Phys. *56*, 2801 (1972)
124. Meyer, W.: Chem. Phys. *17*, 27 (1976)
125. Mulder, F.: Thesis, Nijmegen (1978)
126. Amos, A. T., Yoffe, J. A.: Theoret. Chim. Acta, *42*, 247 (1976)
127. Margenau, H.: Phys. Rev. *38*, 747 (1931)
128. Mulder, F., Thomas, G. F., Meath, W. J.: Mol. Phys., in press; Mulder, F.: unpublished results; Thomas, G. F.: Thesis, London, Canada (1980)
129. Yoffe, J. A.: Chem. Phys. Lett. *61*, 593 (1979); Theoret. Chim. Acta *51*, 207 (1979)
130. Dalgarno, A., Lewis, J. T.: Proc. Phys. Soc. London, *A69*, 57 (1956)
131. Rae, A. I. M.: Mol. Phys. *29*, 467 (1975)
132. Tang, K. T., Toennies, J. P.: J. Chem. Phys. *66*, 1496 (1977) and ref. 30
133. Hepburn, J., Scoles, G., Penco, R.: Chem. Phys. Lett. *36*, 451 (1975) and ref. 29
134. Kreek, H., Meath, W. J.: J. Chem. Phys. *50*, 2289 (1969)
135. Krauss, M., Neumann, D. B.: J. Chem. Phys. *71*, 107 (1979); Krauss, M., Neumann, D. B., Stevens, W. J.: Chem. Phys. Lett. *66*, 29 (1979)
136. Berns, R. M., van der Avoird, A.: J. Chem. Phys. *72*, 6107 (1980)
137. Ng, K. C., Meath, W. J., Allnatt, A. R.: Mol. Phys. *32*, 177 (1976)
138. Ng, K. C., Meath, W. J., Allnatt, A. R.: ibid. *33*, 699 (1977)
139. Cohan, N. V., Coulson, C. A., Jamieson, J. B.: Trans. Faraday Soc. *53*, 582 (1957)
140. Schweig, A.: Chem. Phys. Lett. *1*, 163 (1967)
141. Marchese, F. T., Jaffé, H. H.: Theoret. Chim. Acta *45*, 241 (1977)
142. Holmgren, S. L., Waldman, M., Klemperer, W.: J. Chem. Phys. *69*, 1661 (1978)
143. Howard, B. J.: private communication
144. Tsapline, B., Kutzelnigg, W.: Chem. Phys. Lett. *23*, 173 (1973)
145. Geurts, P. J. M., Wormer, P. E. S., van der Avoird, A.: Chem. Phys. Lett. *35*, 444 (1975)
146. Gallup, G. A.: Mol. Phys. *33*, 943 (1977)
147. Jazunski, M., Kochanski, E., Siegbahn, P.: Mol. Phys. *33*, 139 (1977)
148. Farrar, J. M., Lee, Y. T.: J. Chem. Phys. *57*, 5492 (1972)
149. Monchick, L.: Chem. Phys. Lett. *24*, 91 (1974)
150. Etters, R. D., Danilowicz, R., England, W.: Phys. Rev. *A12*, 2199 (1975)
151. Bauer, W. et al.: Chem. Phys. *17*, 19 (1976)
152. Rulis, A. M., Scoles, G.: Chem. Phys. *25*, 183 (1977)
153. Silvera, I. F., Goldman, V. V.: J. Chem. Phys. *69*, 4209 (1978)
154. Verberne, J. F. C.: Thesis, Nijmegen (1979)
155. Wasiutynski, T., van der Avoird, A., Berns, R. M.: J. Chem. Phys. *69*, 5288 (1978)
156. Scott, T. A.: Phys. Rep. C *27*, 89 (1976) and references therein
157. Raich, J. C., Gillis, N. S.: J. Chem. Phys. *66*, 846 (1977) and references therein

158. Long, C. A., Henderson, G., Ewing, G. E.: Chem. Phys. *2*, 485 (1973); Long, C. A., Ewing, G. E.: J. Chem. Phys. *58*, 4824 (1973)
159. Novick, S. E. et al.: J. Am. Chem. Soc. *95*, 8547 (1973)
160. Rytter, E., Gruen, D. M.: Spectrochim. Acta *35A*, 199 (1979)
161. Gentry, R. W., Hoffbauer, M. A., Giese, C. F.: Abstracts 6 Internat. Symp. Molecular Beams, Trento, Italy (1979)
162. McLaughlin, D. R., Thompson, D. L.: J. Chem. Phys. *70*, 2748 (1979)
163. Kitaigorodsky, A. I.: Molecular Crystals and Molecules, Academic Press, New York (1973)
164. Starr, T. L., Williams, D. E.: Acta Cryst. *A33*, 771 (1977)
165. Warshell, A., Lifson, S.: J. Chem. Phys. *53*, 582 (1970)
166. Filippini, G. et al.: J. Chem. Phys. *59*, 5088 (1973)
167. Huiszoon, C., Mulder, F.: Mol. Phys. *38*, 1497 (1979); *40*, 249 (1980)
168. Van der Avoird, A., Wormer, P. E. S.: Mol. Phys. *33*, 1367 (1977)
169. Van der Linden, J., Van Duijneveldt, F. B.: private communication
170. Van Nes, G. J. H., Vos, A.: Acta Cryst. *B33*, 1653 (1977)
171. Egan, C. J., Kemp, J. D.: J. Am. Chem. Soc. *59*, 1264 (1937)
172. Luty, T., van der Avoird, A., Berns, R. M.: to be published
173. Brith, M., Ron, A.: J. Chem. Phys. *50*, 3053 (1969)
174. Elliot, G. R., Leroi, G. E.: J. Chem. Phys. *59*, 1217 (1973)
175. Luty, T., Van der Avoird, A., Berns, R. M.: J. Chem. Phys., October 1, 1980
176. Steed, J. M., Dixon, T. A., Klemperer, W.: J. Chem. Phys. *70*, 4940 (1979)
177. Sakai, K., Koide, A., Kihara, T.: Chem. Phys. Lett. *47*, 416 (1977)
178. Koide, A., Kihara, T.: Chem. Phys. *5*, 34 (1974)
179. Berns, R. M. et al.: J. Chem. Phys. *69*, 2102 (1978)
180. Whisnant, D. M., Byers Brown, W.: Mol. Phys. *26*, 1105 (1973)
181. Harris, S. J. et al.: J. Chem. Phys. *61*, 193 (1974)
182. Barton, A. E., Chablo, A., Howard, B. J.: Chem. Phys. Lett. *60*, 414 (1979)
183. Mannik, L., Stryland, J. C., Welsh, H. L.: Can J. Phys. *49*, 3056 (1971)
184. Brigot, N. et al.: Chem. Phys. Lett. *49*, 157 (1977)
185. Rich, N. H., McKellar, A. R. W.: Can. J. Phys. *54*, 486 (1976)
186. Bosomworth, D. R., Gush, H. P.: ibid. *43*, 751 (1965)
187. Ryzhov, V. A., Tonkov, M. V.: Opt. Spectrosc. *37*, 476 (1974)
188. Birnbaum, G.: J. Quant. Spectrosc. Radiat. Transfer *19*, 51 (1978)
189. Trafton, L. M.: Astrophys. J. *147*, 765 (1967)
190. Hunt, J. L., Poll, J. D.: Can. J. Phys. *56*, 950 (1978)
191. Van Kranendonk, J.: Physica *23*, 825 (1957); *24*, 347 (1958)
192. Poll, J. D., van Kranendonk, J.: Can. J. Phys. *39*, 189 (1961)
193. Matcha, R. L., Nesbet, R. K.: Phys. Rev. *160*, 72 (1967)
194. Byers Brown, W., Whisnant, D. M.: Mol. Phys. *25*, 1385 (1973)
195. Lacey, A. J., Byers Brown, W.: Mol. Phys. *27*, 1013 (1974)
196. Wormer, P. E. S., van Dijk, G.: J. Chem. Phys. *70*, 5695 (1979)
197. Mazur, P., Jansen, L.: Physica *21*, 208 (1955)
198. Böttcher, C. J. F., Bordewjik, P.: Theory of Electric Polarization, Vol. II, Elsevier, Amsterdam, 1978
199. Gelbart, W. M.: Adv. Chem. Phys. *26*, 1 (1974)
200. Jansen, L., Mazur, P.: Physica *21*, 193 (1955)
201. Orcutt, R. H., Cole, R. H.: J. Chem. Phys. *46*, 697 (1967)
202. Kerr, E. C., Sherman, R. H.: J. Low Temp. Phys. *3*, 451 (1970)
203. O'Brien, E. F. et al.: Phys. Rev. A *8*, 690 (1973)
204. Fortune, P. J., Certain, P. R., Bruch, L. W.: Chem. Phys. Lett. *27*, 233 (1974)
205. Fortune, P. J., Certain, P. R.: J. Chem. Phys. *61*, 2620 (1974)
206. Bruch, L. W., Fortune, P. J., Berman, D. H.: J. Chem. Phys. *61*, 2626 (1974)
207. Dacre, P. D.: Mol. Phys. *36*, 541 (1978)
208. McTague, J. P., Birnbaum, G.: Phys. Rev. Lett. *21*, 661 (1968)
209. McTague, J. P., Birnbaum, G.: Phys. Rev. A *3*, 1376 (1971)
210. Frommhold, L., Proffitt, M. H.: Chem. Phys. Lett. *66*, 210 (1979)
211. Frommhold, L., Proffitt, M. H.: Phys. Rev. A, *21*, 1249 (1980)

212. Proffitt, M. H., Frommhold, L.: Phys. Rev. Lett. *42*, 1473 (1979)
213. Frommhold, L., Proffitt, M. H.: J. Chem. Phys. *70*, 4803 (1979)
214. Oxtoby, D. W., Gelbart, W. M.: Mol. Phys. *29*, 1569 (1975)
215. Oxtoby, D. W.: J. Chem. Phys. *69*, 1184 (1978)
216. Clarke, K. L., Madden, P. A., Buckingham, A. D.: Mol. Phys. *36*, 301 (1978)
217. Sipe, J. E., van Kranendonk, J.: Mol. Phys. *35*, 1579 (1978)
218. Kress, J. W., Kozak, J. J.: J. Chem. Phys. *66*, 4516 (1977)

Van der Waals Systems: Molecular Orbitals, Physical Properties, Thermodynamics of Formation and Reactivity

Pavel Hobza[1] and Rudolf Zahradník[2]

1 Institute of Hygiene and Epidemiology, Centre of Industrial Hygiene and Occupational Diseases, 10042 Prague, Czechoslovakia

2 J. Heyrovský Institute of Physical Chemistry and Electrochemistry, Czechoslovak Academy of Sciences, 12138 Prague, Czechoslovakia

Table of Contents

1 Introduction

1.1 Scope of the Review

Within the frame of this work we have dealt primarily both with quantum chemical calculations on van der Waals (vdW) molecules[1] and with comparison of calculated characteristics with experimental data. In a few cases reference is made to papers in which important experimental characteristics have recently been obtained, which we feel may be of interest to theoreticians. An attempt was made to avoid duplicities and therefore sometimes even very important papers, when easily accessible, are referred to only briefly.

We directed our attention to two broad topics in the very wide subject of vdW interactions, namely on physical properties and chemical reactivities. The former is concerned mainly with various spectroscopic methods. In the latter we considered the reactivity of vdW systems and also their participation in common reactions.

1.2 Classification of van der Waals Systems

Van der Waals molecules may be classified in various ways, two of which will be mentioned here. For physical purposes in general, and particularly for spectroscopic purposes, the classification introduced by Ewing[1)] is valuable. The other possible classification is purely formal[2)] and is based on the number of atoms constituting the subsystems of the vdW system under study. Specifically, for example, the first group comprises systems consisting of a rare gas atom (the first subsystem) and a) a rare gas atom, b) another arbitrary atom (or ion), c) a biatomic molecule (or ion), d) a triatomic molecule, e) any other system (a)–e) is the second subsystem). This classification is used in the subsequent text. The former classification makes, e.g., discussion of the vibrational-rotational spectra of vdW molecules more systematic and logical; the latter should make, e.g., orientation in an extensive table of vdW characteristics easier and more rapid.

Ewing[1)] has distinguished four types of (polyatomic, starting with triatomic) vdW molecules. The first group contains systems in which the interaction of the subsystems is purely isotropic and, therefore, the subsystems rotate freely in the supersystem (free-rotor). The next three groups include systems in which anisotropic interactions operate. The weakest interaction is termed a weak coupling, the next a strong coupling and the product of the strongest interaction (still vdW) is said to possess a semirigid configuration. For triatomic vdW molecules of the $X_2 \ldots Y$ type, the relationship between the magnitude of the leading anisotropic term (in the Legendre polynomial) of the intermolecular potential (describing $X_2 \ldots Y$) and the rotational spacing of the X_2 molecules serves as a classification criterion.

---

[1] The vdW bonds which hold subsystems forming a vdW molecule together are due to permanent or temporary electric multipole—multipole interactions and not, in contrast to common molecules, to electron pair formation.

2 Physical Properties

2.1 Geometry

Calculations of the geometries of molecules, radicals and ions in the electronic ground state and in electronically excited states are among the most important accomplishments of quantum chemistry. Until recently, computation of the geometry was mostly considered in a narrow sense. The system was studied on the basis of chemical experience and intuition and a model was constructed using idealized bond lengths and angles; then only selected bond lengths, bond angles and dihedral angles were optimized. It is true that in some situations this type of partial optimization is much better than no optimization at all, but it is also frequently misleading, and this danger must always be considered.

A physically correct theoretical determination of the molecular structure requires location of minima on the potential energy hypersurface of the system under investigation. In principle, it is possible to achieve this by step-by-step optimization (and reoptimization) of all the internal coordinates. However, this is very tedious even for small systems (consisting of 4 or 5 atoms). Fortunately, in recent years enormous progress has been achieved in locating minima on potential energy hypersurfaces, i.e. of points corresponding to the stable isomers of a given system. It has been found that the most powerful methods for locating minima use the potential energy gradient. Determination of molecular structure is just one aspect of the analysis of potential energy hypersurfaces. The definition, properties and investigation of these hypersurfaces are the subject of Sect. 3.1 where procedures suitable for molecular geometry optimization are also given. Therefore, here we shall give only specific information on the geometry of vdW molecules. First, however, it is important to point out some common features of and especially differences between ordinary and vdW molecules.

First, a careful distinction should be make between the equilibrium and average geometry. The equilibrium geometry is the geometry of the nonvibrating structure corresponding to the minimum on the energy hypersurface. The average geometry characterizes the vibrating molecule and is, in general, not identical with the equilibrium geometry. Although this difference is mostly negligible for ordinary molecules it may be highly significant with vdW molecules. An important feature of the average geometry is that it includes the effect of large amplitude bending vibrations (against the vdW bond), which are typical for vdW systems[1,3].

Second, every vdW systems contains at least one vdW bond, the length of which is much greater than that of any chemical bond and which equals about 0.2–0.5 nm.

Figure 1 schematically depicts the geometries of various vdW molecules determined experimentally and theoretically. With four- and more atomic vdW molecules, theoretical characteristics are given only for structures corresponding to the real minima (vide infra). Wherever possible, an attempt was made to select the best available experimental and theoretical data.

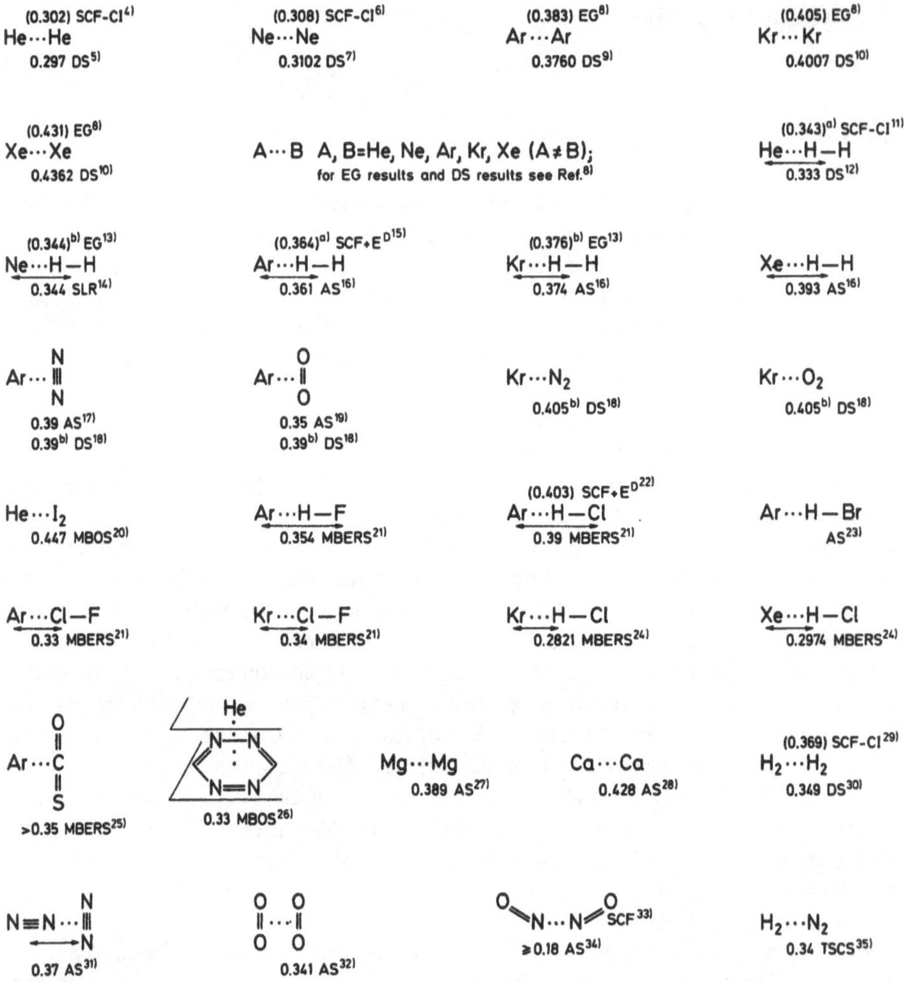

Fig. 1. Experimental and Theoretical Data (in parentheses) on the Structure and Geometry of vdW Molecules.
Theoretical bond lengths and angles are given in parentheses; in some cases only qualitative structural features are available. Nanometers and degrees are used throughout. Abbreviations used: AS — absorption spectroscopy, DS — differential scattering, ED — electron diffraction, EG — electron gas method, MBERS — molecular beam electric resonance spectroscopy, MBOS — molecular beam optical spectroscopy, MS — microwave spectroscopy, SCF — self consistent field method, SCF-CI — self consistent field method with configuration interaction, SCF + E^D — self consistent field method plus dispersion energy, SLR — spin-lattice relaxation, TSCS — total scattering cross-section.
Notes: [a] Calculations indicate that the C_{2v} structure is preferable. [b] Isotropic potential

2.2 Radiofrequency and Microwave Spectroscopy

Klemperer and co-workers[3, 37, 58–61] introduced a very powerful combination of molecular beam techniques with various types of spectroscopy and mass spectroscopy for

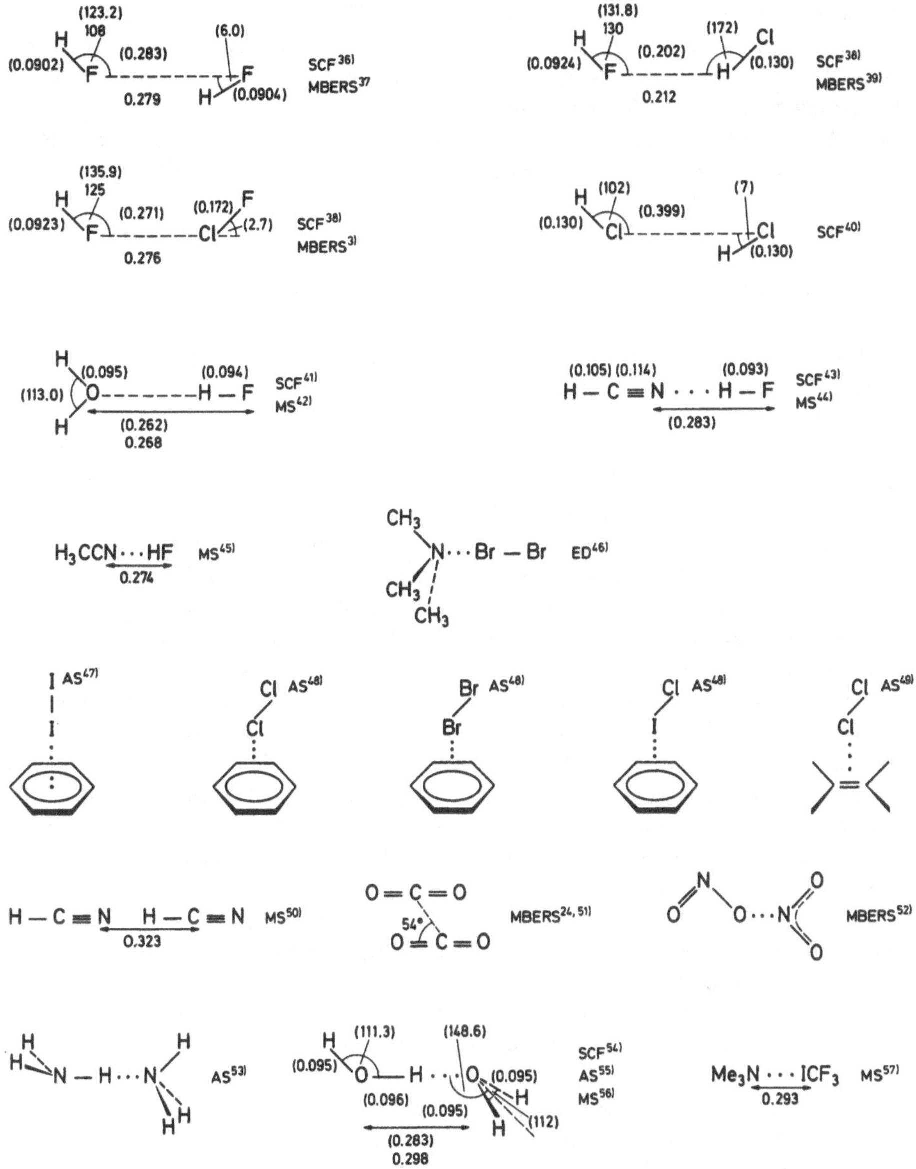

(To figure 1)

investigation of vdW systems. VdW molecules are produced by adiabatically expanding a mixture of subsystems[58] through a supersonic nozzle into an electric resonance spectrometer. In the spectrometer, quadrupole fields are used for state perturbation and analysis. Microwave spectra are recorded in a zero electric field and radio-frequency spectra (nuclear quadrupole resonance) in an external electric field of 1000–2000 V/cm. Finally, the molecular beam enters a mass spectrometer which is used as a beam detector. The HF ... HF[37], Ar ... HCl[59], Ar ... ClF[60] and HF ... ClF[3] vdW molecules were studied in detail. Numerous other attractive vdW systems are mentioned in the review by Klemperer[58] and in a study by Novick et al.[61]. Measurement of the molecular beam electric deflection permitted division of the studied vdW molecules into polar (Ar ... NO, Ar ... HCl, Ne ... DCl, Xe ... HCl, Ar ... BF_3, Kr ... BF_3, NO ... NO, HCl ... HCl, BF_3 ... NO, BF_3 ... CO, CO ... CO) and nonpolar ($(NO)_3$, $(NO)_4$, Ne ... HCl, CO ... H_2, $(BF_3)_n$, n = 2, 3, 4)[61].

2.3 Rotational-Vibrational Spectroscopy and Force Constants

With the exception of molecular beam electric resonance spectroscopy, rotational-vibrational spectroscopy represents the most powerful spectroscopic tool for structure determinations in vdW molecules. Formation of vdW species is manifested in two ways in these spectra. First, the bands characteristic of subsystems are modified slightly and, second, new bands appear due to intersystem vibrations, which are located in the far infrared region (about 15–400 cm^{-1}). This region is not as accessible as the middle infrared region and, therefore, very little data is available on these vibrations. These characteristics are very important for statistical-thermodynamic evaluation of equilibria and reaction rates (for the latter at least within the framework of the absolute rate theory). Low-frequency vibrations are not very important when evaluating zero-point energies but are (at least sometimes) essential for determining the vibrational partition function of the vdW system. Experimental determination of these low frequency vibrations for vdW systems is more important than for common molecules. With common molecules it is rather easy to obtain reasonable estimates of the vibrational characteristics by Wilson FG analysis[62] within the harmonic approximation. The use of this approximation for relatively strong vdW molecules is rather limited (for strong hydrogen bonded complexes of the type polar molecule ... ion ... polar molecule the harmonic approximation even breaks down completely[63]) and it has practically no importance at all for true vdW molecules, as, in these systems, the intersystem vibrations have a strongly anharmonic character (vide infra).

To be able to perform vibrational analysis, we must know the molecular geometry and the complete force constant matrix. These characteristics can be obtained by quantum chemical methods; reliable results can be obtained from nonempirical calculations with at least medium basis sets. In the first step, careful geometry optimization is carried out. Then the quadratic force constants are obtained from the total calculated energies (quadratic fitting to points near the minimum). The G matrix is easily set up from the masses and coordinates of the atoms constituting the molecule under study. In the final step, the eigenvalue problem is solved at the harmonic level. The harmonic frequencies of the normal vibrational modes are obtained by a standard computer program. Details of the procedure are given in numerous books[62,64] and are also described in detail in theoretical studies of the vibrations of hydrogen-bonded complexes[43,54,65]. Comparison of calculated (harmonic) and experimentally observed frequencies is mostly only approximate as the experimental characteristics include vibrational anharmonicities (fundamental frequencies). It is difficult (except with biatomic species) to obtain anharmonicity constants on a comparable quantum chemical level. This problem is the subject of considerable attention. Quantum chemical analysis of vibration anharmonicity of various types of hydrogen bonds was performed by Janoşchek[63]. Moreover, the same author paid much attention to infrared and Raman intensities of various hydrogen-bonded complexes[63]. Sandorfy[63 a)] investigated anharmonicity of hydrogen bonds X—H ... Y by means of analysis of infrared spectra. Experimental anharmonicity for X—H stretching mode and in-plane bending mode are given for alcohols, amines, and thiols. The anharmonicity of the X ... Y bridge vibration is not known.

For vdW biatomics (and vdW quasibiatomics) rather accurate vibrational energy levels for the vdW bond can be obtained by the WKB method[66] when the particular Lennard-Jones (6, 12) potential is available. The WKB eigenvalue equation then assumes the form[67] (for a criticism by le Roy, see Ref.[216]):

$$[2\pi(2\mu)^{1/2}/h]\int_{R_1}^{R_2}[E - D_e - V(R)]^{1/2}\,dR = \left(v + \frac{1}{2}\right)\pi, \tag{1}$$

where μ is the reduced mass of the system and D_e is one of the equilibrium parameters. Low energy levels can readily be obtained with high accuracy by an approximate solution of the eigenvalue equation:

$$E_v/hc = \omega_e\left(v + \frac{1}{2}\right) - \omega_e x_e\left(v + \frac{1}{2}\right)^2, \tag{2}$$

where ω_e and $\omega_e x_e$ are the vibrational and anharmonicity constants, respectively. The following expressions are valid:

$$\omega_e = (6/\pi R_e c)\,(D_e/2\mu)^{1/2}\,, \tag{3}$$

$$\omega_e x_e = (9h/2\mu c)\,(1/\pi R_e)^2\,. \tag{4}$$

The symbols have their usual significance. For the sake of completeness, the Lennard-Jones expression for the vdW bond force constant will be given:

$$f = [d^2V(R)/dR^2]_e = 7.2D_e/R_e^2\,. \tag{5}$$

Investigation of vdW atom-atom complexes has proven rewarding. Much attention has been paid, e.g., to the Ar ... Ar vdW molecule[68]. The potential of the nonrotating ($J = 0$) state contains eight bound vibrational states. There is a continuum of states above the zero-energy level (corresponding to the two argon atoms infinitely far apart). The shape of the potential and the number of bound states are changed if the system is rotationally excited. Then it is necessary to add the centrifugal potential to the potential of the non-rotating vdW molecule, V(R):

$$V(R)^{tot} = V(R) + \frac{h^2}{8\pi^2 c\mu}\,\frac{J(J+1)}{R^2}\,, \tag{6}$$

where J is the angular momentum of the complex and μ is its reduced mass. For a sufficiently high J (e.g., 30)[68] two important changes occur. First, the bound states are now separated by a centrifugal barrier from the two argon atoms at a very large distances. Second, the number of bound states decreases dramatically. Finally, states are created with energy above zero-energy and below the top of the centrifugal barrier; they are termed metastable states. Rotational predissociation of the vdW molecule is conditioned by tunnelling through the centrifugal barrier. Although all these phenomena were discussed[1,69–72] for biatomic vdW molecules they also occur with polyatomic vdW systems.

Collision-induced absorption, which has proven useful for studying vibrational-rotational spectra, was introduced in the fifties by Welsh (for a review, see Ref.[73]). Infrared studies of true vdW molecules in the gas phase require absorption path lengths of up to 200–300 m. This extremely long path is realized by multiple traversals of the radiation through a cell several meters long. This technique (called long path length spectroscopy) is technically complicated but permits recording of very weak features in the fundamental vibration region of the subsystems (i.e. the constituents of the vdW molecule)[70]. This type of spectroscopy has been used, e.g., for the following vdW molecules: H_2 ... H_2 [74], H_2 ... Ar [75], H_2 ... CO [75], Ar ... HX [76] (X is a halogen atom), N_2 ... N_2 [31], O_2 ... O_2 [77], O_2 ... Ar [19], N_2 ... Ar [17].

Matrix isolation vibrational spectroscopy is simpler and easier to apply but is sometimes not as powerful as the previous method. Although not free of various difficulties and shortcomings (in particular, the extent of interactions between the vdW system and the atoms (molecules) of the matrix is not clear), this type of spectroscopy is a useful tool for investigation of hydrogen bonded complexes (self-association and hetero-association) and of molecular (charge-transfer) complexes[53].

In the following part of this Sect. specific information is presented about vibrational characteristics of vdW molecules. The values originate from rotational-vibrational spectra, unless stated otherwise.

The barrier to internal rotation in N_2 ... Ar [17] and O_2 ... Ar [19] is 20 and 30 cm^{-1}, respectively. The WKB stretching vibrational frequency and the anharmonicity constant[67] (in parenthesis) of Ar ... HX (X = F, Cl, Br, I) are 20.0, 15.6, 15.2, 12.0 cm^{-1} (3.80, 1.99, 1.54, 1.11 cm^{-1}), respectively. The heavy atom stretching frequency in the Ar ... HCl system is about 32 cm^{-1} (see the discussion in Ref.[58]). The Ar ... Cl—F bond stretching frequency and the angle bending frequency (the latter in the equilibrium linear structure) are 47 and 41 cm^{-1}, respectively[60].

The vibrational characteristics of $X^{\pm}$ ... H_2O complexes ($X^{\pm}$ is a cation or an anion) were studied theoretically using extended basis sets[78,79] in harmonic approximation. Calculated values for anions are significantly less reliable because the energy hypersurface in the vicinity of the minimum is very flat. Table 1 gives the calculated normal frequencies of these complexes and, for the sake of comparison, the experimental[80] and calculated[78] frequencies for isolated water are also given. The complexes were investigated on two levels of sophistication. In the first (termed a) all force constants were taken from the H.F. potential energy hypersurface of the respective complex, while in the second (termed b), the force constants for the OH stretching and HOH bending were taken from isolated water. The barrier to internal rotation in N_2 ... N_2 [31] is about 15–30 cm^{-1}. The high frequency modes (about

Table 1. Calculated normal frequencies[78,79] (cm^{-1}) for $X^{\pm}$ ... H_2O ($X^{\pm}$ means cation and anion). For comparison experimental[80] and calculated[78] frequencies for isolated H_2O are given

Vibrational mode	$X^{\pm} = Li^+$ a)	Li^+ b)	Na^+ b)	K^+ b)	Mg^{2+} a)	Al^{3+} a)	F^- b)	Cl^- b)	H_2O exp.	H_2O calcd.
ν_1	4052	3715	3714	3714	3929	3420	3715	3714	3714	4032
ν_2	1869	1661	1650	1632	1819	1751	1740	1619	1740	1739
ν_3	445	444	229	176	572	777	292	153		
ν_4	4111	3769	3769	3769	3995	3416	3769	3769	3769	4092
ν_5	521	520	463	398	1047	1322	694	314		
ν_6	529	529	458	444	571	617	1238	759		

[a] All force constants obtained from the H. F. potential energy hypersurface
[b] Force constants for the OH stretching and HOH bending taken from isolated water[80], remaining force constants taken from H.F. potential hypersurface

1555 cm^{-1}) correspond to the in-phase and out-of-phase vibration of the O_2 molecules in the O_2 ... O_2 vdW system[77]. The dimer stretching mode is 23.8 cm^{-1}. Moreover, rotatory modes of the oxygen molecules against the vdW bond were found at 32, 42 and 52 cm^{-1}. These frequencies probably correspond to hindered rotor transitions[77]. In the Raman spectrum of NO ... NO the following low-frequency modes were observed[81] (but not assigned): 167, 196, 262, 478 cm^{-1}. The high frequency modes[82] lie at 1788 and 1860 cm^{-1}. There are four modes of internal motion for the nonlinear structure of HF ... ^{35}ClF (see Section 2.1)[3]: stretching of the vdW bond, in-plane bending of the HF ... Cl angle, two-dimensional bending of the F ... ClF atoms. Assuming that the high frequency covalent stretching motions are independent, and in-plane and out-of-plane bendings are degenerate, the MWS stretching and bending vibrations assume values of 100.5 ± 2 and 170 ± 20 cm^{-1}. The corresponding force constants equal 0.087 mdyne/Å and 0.31 ± 0.08 mdyne Å/rad^2, respectively. The librational mode in CO_2 ... CO_2[83] is 12 cm^{-1}.

Systematic attention has been paid to the vibrational characteristics of hydrogen bonded complexes. No experimental vibrational data are available for HF ... HF but thorough non-empirical studies have been published[36,65]. The frequencies calculated in extended and 4-31G basis sets are given in Table 2. Frequencies were also

Table 2. Normal frequencies[36,65] for HF ... HF and HF calculated in 4-31G and extended basis sets (cm^{-1})

Vibrational mode	HF ... HF		HF[a]	
	4-31G	extended[b]	4-31G	extended[b]
ν_1	4081	4416	4117	4450
ν_2	4038	4364	—	—
ν_2	588	544	—	—
ν_4	519	208	—	—
ν_5	226	146	—	—
ν_6	171	140	—	—

[a] Exp.: $\nu = 4139$ cm^{-1} (Ref.[65])
[b] Calculation for this review based on force constants from Ref. [36]

calculated for FD ... FH, FH ... FD and FD ... FD[65]. Sharp absorption due to HCl ... HCl was observed among the rotational-vibrational lines of the monomer[84]. The results of a nonempirical SCF study (4-31G) are given in Ref.[40]. The H_2O ... HF complex was the subject of infrared[85], microwave[42,86] and nonempirical SCF studies (6-31G)[41]. The HCN ... HF complex was also studied[43,86,87] intensely. Calculated and experimental frequencies are compared in Table 3. Differences between calculated and observed frequencies were ascribed to the fact that the theoretical values are harmonic frequencies while the experimental values are fundamental frequencies. In the cited work[43] the experimentally observed changes in intensity associated with hydrogen bond formation were interpreted quantum chemically. The experimentally determined intermolecular stretching frequencies[88] of $(CH_3)_2O$ ... HF, $CH_3C_2H_5O$ ... HF and $(C_2H_5)_2O$ ... HF equal 185, 180 and 175 cm^{-1}; the intermolecular stretching frequency[89] in $(CH_3)_2O$ ... HCl has a value of 119 $\pm$ 4 cm^{-1}. Intermolecular vibrations of CH_3CN ... HF were investigated experimentally (infrared[87] and microwave[45] spectroscopy) and theoretically (4-31G)[90]. The predicted hydrogen bond stretching (191 cm^{-1}) and bending (52 cm^{-1}) agree well with the respective experimental data (168, 40[87]; 181, 45[45]). The next systematically investigated complex was the water dimer[54,56,91]. The 4-31G basis set gives too strong a hydrogen bond which is manifested in too large values for some force constants. The frequencies of the intermolecular vibrations qualitatively fit the infrared spectrum of liquid water[54]. It has been shown in a study on hexagonal ice[91] that inclusion of the correlation energy improves both the calculated geometry and the calculated frequency values. The six calculated intermolecular frequencies are of three types: (a) two hydrogen bends (about 400–500 cm^{-1}), (b) a hydrogen bond stretch (about 200 cm^{-1}) and (c) three low-frequency vibrational motions (about 80 to 180 cm^{-1}) in which the linear hydrogen bond is not involved[54]. However, with the water dimer the H.F. geometry fits better microwave geometry found by Dyke and Muenter[56].

Self-association of HCN was studied by infrared[92] and microwave spectroscopies[50] and it was concluded that the dimer is linear. Matrix isolation studies support this

Table 3. Calculated and experimental normal frequencies (cm^{-1}) for the HCN ... HF and H_2O ... HF complexes

Vibrational mode	HCN ... HF		H_2O ... HF	
	4-31G[a]	exp.[b]	6-31G[c]	exp.[d]
ν_1	3990	3710	4122	
ν_2	3682		3974	
ν_3	2396		3747	3608
ν_4	937 }		1755	1600
ν_5	937 }		913	696
ν_6	561 }	555	740	666
ν_7	561 }		276	198
ν_8	193	155	265	94
ν_9	86 }	70	250	180
ν_{10}	86 }		—	—

[a] Ref.[43], [b] Ref,[87], [c] Ref.[41], [d] Ref.[42]

conclusion[53,93–95]. In Ref.[53] trimers and tetramers were also studied. The calculated force constants indicate that the additional hydrogen bond in the trimer is stronger than the first hydrogen bond in the trimer and than that in the dimer. This is a result of charge redistribution which was also predicted by nonempirical quantum chemical calculations[96].

For numerous ion ... polar molecule hydrogen bonded complexes harmonic and anharmonic vibrational frequencies were calculated and compared with observed ones; for the respective references see Ref.[63]. A detailed description of calculating vibrational spectra and infrared absorption intensities is given in a study on the $(CH_3SH)_2$ dimer[97].

Numerous papers deal with the vibrational spectroscopy of charge-transfer complexes. Only a few recent papers are mentioned here because they deal with systems for which the possibility of performing deeper theoretical investigations in near future is very promising. The infrared spectra of the following complexes were studied in nitrogen (and argon) matrices: H_2O ... CO_2[98] (for an ab initio study see Ref.[99]) H_3N ... CO_2[100], H_3N ... Cl_2[101], H_2O ... H_2O[55], benzene ... Cl_2, Br_2, ICl[48], benzene ... I_2[47], H_2S ... CH_2O[102].

2.4 Electronic Spectroscopy

While the energy well of the ground state of vdW molecules is always very shallow, this need not be true of the first electronically excited state. With the singlet ground state vdW molecule, both the singlet and triplet excited states can exhibit deepening of the energy well. A similar sort of deepening of the energy well occurs when a vdW molecule loses an electron. Such a process occurs in photoelectron spectroscopy and the resulting radical-cation is much more stable than the dissociation products, i.e. the free molecule and the radical cation (see Sect. 2.5).

The first electronically singly excited state of a vdW molecule may be regarded as a complex of an electronically excited subsystem and a ground state subsystem. Complexes of this sort have been known for years and are called excimers. Excimers derived from rare gas atoms and from closed shell hydrocarbons have been studied most frequently. Excimers are sometimes rather easily obtained through interaction of two excited triplet states of the subsystem. As a result of this triplet-triplet annihilation a molecular pair is formed, consisting of a ground state subsystem and a first excited singlet subsystem, i.e. the excimer. The emission spectra of numerous excimers have been recorded and analyzed[103,104]. It is impossible to ascribe the excitation to just one of these subsystems; for the first order configuration interaction the wavefunctions of the two excited states formed are given by the relationships (identical subsystems):

$$\Psi_{1,2} = \frac{1}{\sqrt{2}} (\Psi^R_g \Psi^S_e \pm \Psi^R_e \Psi^S_g) \tag{7}$$

where Ψ^R and Ψ^S are the wave functions of subsystems R and S and g and e are subscripts designating the ground and excited states. The situation which does not permit localization of the excitation on a certain system is termed excitation delocalization[105].

Murrell[106] described and illustrated the theory of weakly interacting chromophores. His approach may be used in a straightforward way for analysis of the electronic spectra of vdW systems. Unless the states of one subsystem are mixed by the field of the other subsystems (i.e. in the absence of the field effect), the decisive matrix elements of the total Hamiltonian between the ground state, the singly excited states and the doubly excited state are then (the symbols used by Murrell[106] are employed):

$$\langle R_m(i)\, S_n(j)|\, \hat{H}'(i, j)\, |R_p(i)\, S_q(j)\rangle\,, \tag{8}$$

where R_m and R_p are the excited states of the first subsystem (whose ground state is R_0), S has a similar meaning for the second subsystem and $m \neq p$, $n \neq q$. $\hat{H}'(i, j)$ is the interaction Hamiltonian describing electronic repulsion between electron set i (of the first subsystem) and electron set j (of the second subsystem). If the number of electrons of the first and second subsystem equal n_i and n_j, respectively, integral can be expressed as a n_in_j product of a representative contribution (including, e.g., the 1st and the 2nd electron):

$$n_in_j\langle R_m(i)\, S_n(j)|\, e^2/4\pi\varepsilon_0 r_{12}\, |R_p(i)\, S_q(j)\rangle\,. \tag{9}$$

Integration of (9) leads to

$$\int \varrho^R_{mp}(1)\, \varrho^S_{nq}(2)\, (e^2/4\pi\varepsilon_0 r_{12})\, dx_1\, dx_2\,, \tag{10}$$

where the transition densities, ϱ, are defined by

$$\varrho^R_{mp}(1) = n_i \int R^*_m(i)\, R_p(i)\, d\tau\,, \tag{11}$$

$$\varrho^S_{nq}(2) = n_j \int S^*_n(j)\, S_q(j)\, d\tau\,. \tag{12}$$

In the integrations in (11) and (12) the position coordinates of electrons 1 and 2 are not included. The dipole moments associated with the transition densities are the transition (dipole) moments for the excitations within the individual subsystems ($R: m \rightarrow p$; $S: n \rightarrow q$). The interaction energy between excited states R_mS_n and R_pS_q, represented by integral (9), is the electrostatic energy of the two transition densities, one located on R and the other on S.

When interpreting integral (9), Murrell[106] demonstrated that interaction between the ground state (R_0S_0) and the doubly excited state (R_mS_n) leads to London stabilization of the ground state (the dispersion energy). When investigating the electronic spectra of vdW molecules, interaction between singly excited states (R_0S_n and R_mS_0) is most interesting. When considering the splitting of the excited state, it is necessary to evaluate the interaction matrix element which is connected with the interaction energy of the respective transition densities, ϱ^R_{0m} and ϱ^S_{0n}. The split between the two excited states is given by twice the interaction matrix element. Murrell[106] illustrated this result by two orientations of the transition dipole moment in the ethylene molecules (Table 4). He was able to interpret the shift of the absorption band (which is due to the C=C bonds) observed when passing from [2,2,1]-bicycloheptene to [2,2,1]-bicycloheptadiene. For symmetry reasons, transitions into the in-phase state (the upper line in Table 4) are allowed and those into the out-of-phase state are forbidden. Thus the allowed band of the supersystem is either bathochromically (configuration 1) or hypsochromically (configuration 2) shifted with respect to the band of the subsystem. In conclusion it should be noted that the split of the excited states is proportional to the dipole-dipole interaction energy only when transitions in both subsystems are symmetry allowed. If not, then the dipole-higher multipole or the higher multipole-higher multipole assumes the crucial role.

Table 4. Spectral characteristics of two interacting ethylenes[a] (according to Murrell[106)])

Interaction energy		Wave function	Energy	
config. 1	config. 2		config. 1	config. 2
$-2\mu^2/r^3$	μ^2/r^3	$\frac{1}{\sqrt{2}}(R_N S_V + R_V S_N)$	$E_V - 2\mu^2/r^3$	$E_V + \mu^2/r^2$
		$\frac{1}{\sqrt{2}}(R_N S_V - R_V S_N)$	$E_V + 2\mu^2/r^3$	$E_V - \mu^2/r^2$

[a] Configuration 1: 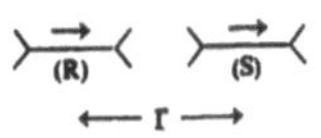

Configuration 2: 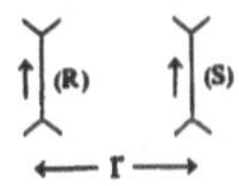

The arrows indicate the orientation of the transition dipole moments

We have used[107)] semiempirical CNDO/S calculations[108)] for estimating the magnitude and nature of shifts accompanying the transition from a molecule to its vdW dimer. $(H_2)_2$, $(N_2)_2$, $(HF)_2$, $(HCl)_2$, HCl ... HF, $(H_2O)_2$ and $(CO_2)_2$ were studied. Qualitatively speaking, the results agree with those obtained by the Murrell approach. Specifically, e.g., with the $(H_2)_2$ vdW molecule, the four most frequently studied configurations (designated 1, 3, 6, 8 in Table 5) were analyzed and it was found that no shift is to be expected with complexes 3 and 8 (for symmetry reasons) and a larger split should be connected with 1 ($\sim 1500\ cm^{-1}$) than with 6 ($\sim 700\ cm^{-1}$). Coulomb energies corresponding to dipole and (linear) quadrupole interactions are given in Table 5. They can be used for estimates of important features of the electronic spectra of vdW molecules.

Table 5 is applicable to real vdW molecules alone but it can also be used for "vdW units" incorporated into ordinary molecules. The [1,3,3]-bicycloheptadiene molecule, mentioned above, belongs in this class. In other words, weakly interacting structural units incorporated into a complex molecular frame could serve

Table 5. Coulomb interaction energies between two (transition) dipoles (μ—μ), dipole — linear quadrupole (μ—Q) and two linear quadrupoles (Q—Q)

No.	Configuration[a]		E^C		
	R	S	μ—μ	μ—Q	Q—Q
1	→	→	$-\frac{2\mu_R\mu_S}{r^3}$	$\frac{3\mu_R Q_S}{r^4}$	$\frac{24}{16}\frac{Q_R Q_S}{r^5}$
2	→	↗	$-\frac{\mu_R\mu_S}{r^3}$	$\frac{3}{4}\frac{\mu_R Q_S}{r^4}$	$-\frac{3}{16}\frac{Q_R Q_S}{r^5}$
3	→	↑	0	$-\frac{3}{2}\frac{\mu_R Q_S}{r^4}$	$-\frac{12}{16}\frac{Q_R Q_S}{r^5}$
4	→	↖	$\frac{\mu_R\mu_S}{r^3}$	$\frac{3}{4}\frac{\mu_R Q_S}{r^4}$	$-\frac{3}{16}\frac{Q_R Q_S}{r^5}$
5	→	←	$\frac{2\mu_R\mu_S}{r^3}$	$\frac{3\mu_R Q_S}{r^4}$	$\frac{24}{16}\frac{Q_R Q_S}{r^5}$
6	↑	↑	$\frac{\mu_R\mu_S}{r^3}$	0	$\frac{9}{16}\frac{Q_R Q_S}{r^5}$
7	↑	↓	$-\frac{\mu_R\mu_S}{r^3}$	0	$\frac{9}{16}\frac{Q_R Q_S}{r^5}$
8	|		0	0	$\frac{3}{16}\frac{Q_R Q_S}{r^5}$

[a] The angle between the dipole moment axes in cases 2 and 4 is 45° and 135°, respectively

as very useful examples of rather rigid vdW systems. Numerous non-bonding transannular interactions belong in this group. Althought the electronic spectra of many of these systems have been successfully interpreted, it might be rewarding to try to reinterpret their spectra considering the spectroscopy of vdW molecules.

Systematic and extensive attention has been paid to the high resolution electronic (vacuum ultraviolet) absorption spectra of X ... X[109–114] and X ... Y[110,113] molecules, where X, Y are noble gas atoms. Measurements were performed in the 58–126 nm region with a 6.65 m vacuum spectrograph. The individual band systems were identified and very accurate characteristics were obtained for the ground states (the potential depths, number of stable vibrational levels, rotational constants and interatomic separations).

The absorption band of $O_2 \ldots O_2$ was assigned to collision-induced absorption[77,115]. Moreover, discrete features were found which certainly correspond to bound state $O_2 \ldots O_2$ molecules. The dissociation energies of both the ground and electronically excited states were obtained from analysis of the dimer vibrational levels. The upper electronic state was identified as a combined state $^1\Delta_g$ (v = 0) + $^1\Delta_g$ (v = 1); the ground state is $^3\Sigma_g^-$ (v = 0). All the observed dimer features are shifted to higher frequencies (17278–17321 cm^{-1}) with respect to the isolated double molecule transition (17248 cm^{-1}).

The electronic spectra of the NO ... NO[82], $Br_2 \ldots Br_2$[116] and $I_2 \ldots I_2$[117,118] dimers were used for studying dimerization equilibria. The absorption band maxima are located at 37200 and 40800 cm^{-1} with I_2 and $I_2 \ldots I_2$, respectively.

The relatively small energy gap between the two lowest excited states of vdW molecules (vide supra) suggests that $T_1 \rightarrow T_x$ and $S_1 \rightarrow S_x$ transitions could be located in the near infrared or visible regions. A recent near infrared study[119] of electronically excited dimers (Ne ... Ne*, Ar ... Ar*, Kr ... Kr*, Xe ... Xe*) supports this tempting idea. Several heteronuclear ions, XY^+, (X, Y = He, Ne, Ar, Kr) were investigated[120,121]. It should be noted that, e.g., the first two band groups with the $HeNe^+$ cation are located in the visible region (425 and 410 nm)[121].

In concluding this Sect. on classical electronic spectroscopy, it should be mentioned that weak absorption bands of vdW systems are frequently partly overlapped or completely hidden by the strong bands of the subsystems. Nonetheless there are still quite a few attractive systems worth investigating by this method.

A few years ago, a powerful version of molecular optical spectroscopy with supersonic beams and jets was developed by Smalley, Wharton and Levy[26]. Supersonic expansion of molecules in an inert carrier gas yields an ideal spectroscopic sample. As a result of the expansion, the translational temperature of the carrier gas decreases to extremely low values (below 0.1 K). The flow is collisionless so that even extremely unstable species survive. Special attention was paid[26] to fluorescence excitation spectroscopy but the technique is by no means limited to this type of spectroscopy. (Because of fundamental difficulties, however, direct measurement of light absorption in molecular beams is not easy.) Cooled molecules in the beam are electronically excited with a tunable dye laser. The emitted fluorescence is detected and plotted against the wavenumber of the exciting radiation. The obtained fluorescence excitation spectrum is generally very similar to the corresponding absorption spectrum. The technique was used for analysis of the spectra of interesting vdW molecules: He ... NO_2, He ... I_2, X ... tetrazine and X_2 ... tetrazine (X = He, Ar, H_2) complexes[26].

2.5 Electron Spectroscopy (PES, PIES, ESCA) and Ionization Potentials

Electron spectroscopic techniques are based on determination of the energy distribution of electrons released in the ionization process. Two of these techniques became very popular among chemists and molecular physicists, namely photoelectron spectroscopy (PES)[122] and X-ray electron spectroscopy also termed electron spectroscopy for chemical analysis (ESCA)[123]. Penning ionization electron spectroscopy (PIES)[124] is related to PES, but the target molecule is ionized by electronically excited (metastable) atoms of a noble gas, mostly He, Ne and Ar instead of the photons used in PES. PIES is not such a widely used technique as PES and ESCA, but probably the most attractive one for vdW molecules.

First a short account on PES and photoionization studies will be given. A large shift was observed[125,126] in ionization potentials when passing from a closed shell system to its vdW dimer. This striking shift can be qualitatively understood by considering that the state after ionization is, in general, more strongly bonding than the initial state. Clearly, there is some resemblance between the bonding properties of the radical cation derived from a vdW dimer and an excimer.

The cited papers[125,126] deal with the dimers of rare gas atoms but extension to other vdW molecules can be expected in the near future. The importance of combining the PES technique with supersonic molecular beams has been pointed out: in Ref.[125] in Xe ... Xe production a jet with a 35 µm aperture was used. No peak was observed at about 11.1–11.2 eV (the 1st adiabatic potential of Xe ... Xe is 11.14 eV[126]) but a small gradual intensity increase culminates at about 11.7 eV (a small, flat maximum) which was tentatively attributed to a transition to an excited vibrational level of the ionic ground state (Xe_2^+ : $A^2\Sigma^+_{(1/2)u}$). Further Xe ... Xe peaks were observed at 11.85 ($B^2\Pi_{(3/2)g}$), 12.0 ($B^2\Pi_{(1/2)g}$), 12.2 ($C^2\Pi_{(3/2)u}$), 13.2 ($C^2\Pi_{(1/2)u}$) and 13.3 ($D^2\Sigma^+_{(1/2)g}$) eV.

Photoionization studies of rare gas dimers have been carried out[126–128]. Photoionization efficiency curves for the vdW molecules (Ar ... Ar, Kr ... Kr, Xe ... Xe) were obtained by means of molecular beam techniques. The dissociation energies of the ions were calculated from the measured first ionization potentials (Table 6) and the dissociation energies of the parent vdW molecules. As the dissociation energy of vdW molecules is negligible compared the dissociation energy of the corresponding cations, the latter energies (available from other sources) can be used to estimate the first ionization potentials of vdW systems which have not yet been studied (values in parentheses in Table 6).

Analysis of photon yield curves of rare gas dimers has shown that the main mechanism participating in photoionization is autoionization[126].

A simplified scheme describing Penning ionization can be written:

$$^{1,3}N^* + {}^1A \rightarrow {}^2A^{\oplus} + {}^1N + e\,, \tag{13}$$

where $^{1,3}N^*$ is an electronically excited rare gas atom in the singlet or triplet state, ^1A is the target molecule (in its singlet ground state) and e is an electron.

The PES and PIES spectra are similar but the corresponding peaks are mostly shifted either to larger or to smaller electron energies[130–133]. These shifts are probably caused by interaction either between the target molecule and the rare gas atom in an excited state or between the ion produced and the rare gas atom in the ground state. The ionization itself involves a vertical transition from the upper potential surface of the reactants ($N^* + A$) to the lower surface of the products

Table 6. First ionization potential of vdW molecules: Rare gas dimers. For comparison are included also potentials for atoms (Ref.[129], unless otherwise stated)

System	I (eV)	System	I (eV)
He	24.581	Kr	13.996
He ... He	(21.5)[a]	Kr ... Kr	12.87
Ne	21.559	Xe	12.127
Ne ... Ne	(20.2)[a]	Xe ... Xe	11.14
Ar	15.755	Rn	10.746
Ar ... Ar	14.54	Rn ... Rn	(~10.0)[b]

[a] An estimate based on the dissociation energy of X_2^+

[b] An estimate based on extrapolation

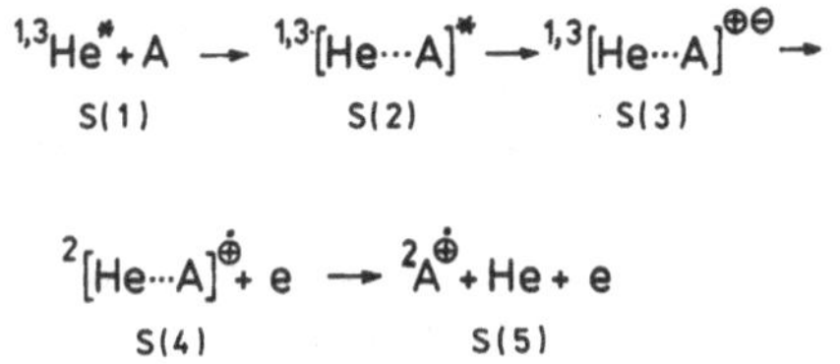

Fig. 2. Mechanism of Penning Ionization. (Taken from Ref.[2])

($\dot{A}^{\oplus}$ + N). Under certain conditions, the minimum depth on one of the two potential surfaces can be determined from the peak broadening and shape[131].

The ionization process can be described as follows (assuming that N = He; Fig. 2): At stage S(1) the components are far apart; at S(2) a collision complex can be formed between the excited helium atom and molecule A, and in the following stage, S(3), an electron is transferred from A to the 1s orbital of He* and an electron is then released from the 2s orbital (stage S(4)). The ^{2}[He ... A]$^{\oplus}$ complex is short-lived and decomposes to $^2A^{\oplus}$ and He (stage S(5)). Stages S(2) and S(4) are especially interesting from the point of view of weak intermolecular interactions. To understand the differences between the peak positions on the energy scale in the PES and PIES spectra, it is necessary to study theoretically the nature of the complexes in stages S(2) and S(4). This is still a difficult task, as knowledge on the interactions of excited systems with the ground-state systems (stage S(2)) is fragmentary; experience with the interactions of two open-shell systems (stage S(3)) is also limited.

The photoelectron spectra of hydrogen-bonded complexes (between fluorocarboxylic acids and trialkylamines) were recorded in the gas phase[134]. The non-bonding orbital of trialkylamines is stabilized significantly by hydrogen-bond formation. Photoelectron spectroscopy has also been used for studying intramolecularly hydrogen-bonded systems[135]. The pyridine-iodomonochloride charge-transfer complex has been studied in the gas phase by ESCA[136]. The atomic core energy shifts observed suggest a transfer of 0.1 electron from the nitrogen to the iodine atom of the complex.

The vertical ionization potential of $(H_2)_n$ clusters (n = 2–8) were calculated[137] by the ab initio SCF method (Koopmans' theorem). The change in the correlation energy on ionization was found from the difference in the semiempirical (EPCE-F2σ) correlation energies of the parent closed shell cluster and its cation, assuming the same geometry for both species. The effect of the reorganization energy on the ionization potential was ignored. With the clusters investigated we found a decrease in the vertical ionization potential by about 0.4–0.6 eV (compared to the H_2 molecule).

3 Thermodynamics of Formation of van der Waals Molecules

3.1 Potential Energy Hypersurfaces and Their Stationary Points

The potential energy of a system is a function of all its internal coordinates. This function is usually interpreted geometrically as a surface in n-dimensional configurational space, called an energy hypersurface. With a diatomic molecule, described by a single internal coordinate, the hypersurface simplifies to a potential energy curve. Whereas an experienced chemist dealing with classical molecules is usually able to assess whether a given isomer corresponds to a minimum or to a saddle point, the situation is much more complicated with vdW systems. Hence it is desirable not only to locate minima on energy hypersurfaces and to investigate their properties but also to search for saddle points on the hypersurface; this process is of basic importance for the investigation of vdW reactivity.

The points on an energy hypersurface, for which the first derivatives of the total energy with respect to all the coordinates of all the atoms are equal to zero, are termed stationary points. Thus, it must hold that

$$\frac{\partial E}{\partial x_1} = \frac{\partial E}{\partial x_2} = \dots \frac{\partial E}{\partial x_n} = 0\,, \tag{14}$$

where E is the total energy of the complex and x_i is the set of coordinates for atom i. In determining the character of a stationary point, the second derivatives of the energy with respect to all the coordinates (force constants) must be known[138)]. The matrix of force constants is diagonalized[62)] and the nature of the individual stationary point is found on the basis of the number of negative eigenvalues[139)]. If all the eigenvalues are positive, then a local minimum is involved, corresponding to a stable conformation of the vdW molecule. If one eigenvalue is negative, a saddle point is present, corresponding to an activated complex in the sense of the theory of absolute reaction rates. The eigenvector corresponding to the negative eigenvalue yields information on the direction of the reaction coordinate[140,141)]. If the force constant matrix has more than one negative eigenvalue, then the stationary points correspond neither to minima nor to saddle points.

Location of the stationary points on a vdW potential energy hypersurface represents a rather complicated task. The classical approach, in which all internal coordinates of the complex are determined by step-by-step optimization (and reoptimization), is tedious and time consuming. This approach is, in practice, limited to vdW molecules with no more than 6–8 atoms (corresponding to 12–18 internal coordinates with nonlinear complexes). More powerful methods for locating stationary points on potential energy hypersurfaces use the gradient of the total energy[142)]. Within semiempirical methods of the CNDO type, very efficient programs were developed by McIver and Komornicki[143)] and by Pancíř[144)]. In the latter program, combination of the Murtagh and Sargent procedure[145)] for calculating the energy gradient with a double-iteration technique (i.e. with geometry optimization in each iteration of the SCF procedure) yielded reliable results with a minimum of effort: complete geometry optimization lasts only 200–300% longer than calculation of the energy for a single hypersurface point. In connection with the ab initio SCF method, the very efficient TEXAS program, developed recently by Pulay[146)] should be mentioned. The respective ab initio gradient program is characterized by efficiency of the gradient evaluation and by the ability to handle higher angular momentum (d and f) basis functions. The computing time for force calculation is not greater than 500% of that necessary for a single SCF calculation.

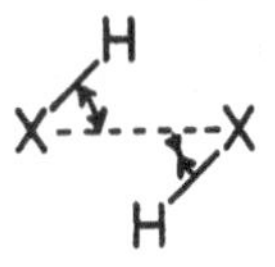

Fig. 3. Structure of the Saddle Points of $(HF)_2$ and $(HCl)_2$ Dimers[40,65]; X = F, Cl

Table 7. Procedures for location of stationary points on energy hypersurface

Method	References
Empirical potential	147–150
EHT	151
CNDO/2	143, 144, 152
INDO	144
MINDO/2	144
MNDO	153
Ab initio SCF (lobe functions)	154–156
Ab initio SCF	146, 157–160
Ab initio SCF-CI	161

References to various procedures for location of stationary points on energy hypersurface are listed in Table 7 for easier orientation.

The geometry characteristics of vdW molecules with real minima are given in Fig. 1. It should be mentioned that the number of polyatomic (four and more atomic) vdW molecules investigated on the ab initio SCF level (complete vibrational analysis) is still very limited: $(HF)_2$, $(HCl)_2$, HF ... HCl, HF ... FCl, HCN ... HF, H_2O ... HF, $(H_2O)_2$ (for references see Fig. 1). The structure of the saddle points of $(HF)_2$ and $(HCl)_2$[40,65] is depicted in Fig. 3 and the vibration (corresponding to the negative eigenvalue) which changes the saddle points into real minima is indicated. Recently, Kerns and Allen[162] investigated the cyclic structures of $(H_2O)_2$, $(HF)_2$, $(H_2S)_2$ and $(HCl)_2$ and the bifurcated structures of H_3N ... H_2O, $(H_2O)_2$, HF ... H_2O and $(H_2S)_2$ (ab initio SCF with a 4-31G basis set). The authors[162] have shown that cyclic and bifurcated structures correspond to saddle points, not to real minima. However, this finding was not based on complete vibrational analysis, but on investigation only one internal coordinate.

In order to make the time-consuming ab initio analysis of parts of energy hypersurfaces with polyatomic systems less tedious, an attempt was made to use CNDO/2 calculations for preliminary estimations of the structure and nature of saddle points. The following complexes were studied: $(H_2O)_2$, $(HF)_2$, $(LiF)_2$, CH_3OH ... HF, $(CH_3OH)_2$[163], CH_2O ... H_2O[164].

3.2 Statistical Thermodynamic Treatment and the Role of Entropy

The process leading to the formation of a vdW molecule, R ... T, will be investigated in some detail:

$$R + T \rightleftharpoons R \dots T\,. \tag{15}$$

The equilibrium constant (K) of the process at temperature T is related to the standard change in the Gibbs energy, ΔG^0 (16):

$$\Delta G_T^0 = -RT \ln K_T . \tag{16}$$

The ΔG^0 term can be determined from knowledge of the enthalpy and entropy of complex formation, ΔH^0 and ΔS^0, using the usual equation:

$$\Delta G_T^0 = \Delta H_T^0 - T\,\Delta S_T^0 . \tag{17}$$

With strong interactions, i.e. with chemical processes, the absolute value of the ΔH^0 term is relatively large, and, therefore, ΔG^0 is essentially given by ΔH^0. The absolute value of ΔH^0 accompanying the formation of vdW systems is much smaller; the entropy term can then play a more important role. It is preferable to deal separately with equilibria involving true vdW systems and stronger complexes (e.g., hydrogen bonded systems, hydrated ions). The latter group will be discussed first.

In order to calculate the thermodynamic functions of the process described by Eq. (15), it is necessary to known the equilibrium geometry and the frequencies of the normal vibrational modes of all species involved in the equilibrium process, as well as interaction energy, ΔE. Partition functions, used for relatively strong vdW molecules, were evaluated using the rigid rotor-harmonic oscillator approximation.

Table 8 contains the thermodynamic characteristics for the formation of various types of complexes, starting with a strong dipole-dipole complex, followed by different hydrogen-bonded complexes and ending with a true vdW molecule. For most complexes, the equilibrium geometry, harmonic frequencies, and ΔE were determined using ab initio SCF calculations in the 4-31G basis set[38,40,43,54,65]. Only for $(LiH)_2$[166] and $H_2O \ldots HF$[41] were other basis sets used (Li: [4s2p]; H: [2s1p] and 6-31G, respectively). For the Ar ... ClF complex, the input values were based on experimental data[60]. Experimental values (mostly incomplete) are available only for a few complexes[84,85,167,168]. With $H_2O \ldots HF$ and $(H_2O)_2$, the usual overestimation of the n-31G values is observed; otherwise, the agreement is satisfactory, which is also true of the two available entropy values.

Analysis of the entries in Table 8 allows some useful conclusions to be drawn. The ΔH_0^0 term differs from ΔE by the zero-point vibrational energy, which frequently attains relatively large values (e.g., 20% of ΔE). The magnitude of the zero vibrational energy is understandable in the light of the fact that the force constants were evaluated using the SCF calculation with a medium basis set. If the vibrational frequencies were evaluated on the H.F. level (or even behind this level), their values and their zero-point vibrational energies become smaller (vide infra). Not only is the difference between ΔE and ΔH_0^0 important, but the temperature dependence of ΔH^0 also plays a significant role. The difference between ΔH_0^0 and ΔH_{298}^0 frequently represents more than 10% of the ΔH_0^0 value. These aspects were also analyzed[169] in connection with a study on the isomerism and equilibrium behaviour of the NO ... NO complex. Thermodynamic analysis was per-

Table 8. Thermodynamic characteristics for the formation (kJ/mol) of various types of van der Waals molecules (taken from Ref.[165]) unless otherwise stated). Experimental results are given in parentheses

Reaction	ΔE	ΔH_0^0	T(K)	ΔH^0	$T\Delta S^0$	ΔG^0	K
$LiH + LiH \rightleftharpoons (LiH)_2$[a]	−197.48		300	−185.64	−41.96	−143.68	1.03×10^{25}
			500	−186.02	−71.42	−114.60	9.38×10^{11}
$H_2O + HF \rightleftharpoons H_2O \ldots HF$	− 56.47	−44.79	100	− 47.36	− 9.70	− 37.66	4.7×10^{19}
	(−30 ± 7)[b]		298.2	− 49.27	−32.61	− 16.67	831.8
			315	− 49.30	−34.48	− 14.83	287.7
				(−26 ± 5)[b]	(−28 ± 2.5)[b]		(0.24)[b]
			500	− 48.86	−54.23	5.37	0.27
$HCN + HF \rightleftharpoons HCN \ldots HF$	− 38.38	−29.94	100	− 31.77	− 9.75	− 22.02	3.2×10^{11}
			298.2	− 32.10	−30.01	− 2.09	2.3
			500	− 30.72	−48.61	17.89	0.01
$H_2O + H_2O \rightleftharpoons (H_2O)_2$	− 34.39	−24.28	100	− 26.46	− 8.22	− 18.24	3.38×10^{9}
			298.2	− 26.06	−24.40	− 1.66	1.95
			376	− 25.26	−29.87	4.61	0.23
				(−15.19 ± 2.1)[c]	(−29.28 ± 2.0)[c]		(0.011)[c]
			500	− 23.72	−37.97	14.25	0.03
$HF + HF \rightleftharpoons (HF)_2$	− 33.47	−25.17	100	− 27.19	− 7.92	− 19.27	1.2×10^{10}
	(−29.3)[d]		298.2	− 28.48	−26.22	− 2.27	2.5
			500	− 27.92	−43.29	15.37	0.02
$HF + HCl \rightleftharpoons HF \ldots HCl$[e]	− 19.55	−13.53	100	− 15.34	− 7.94	− 7.40	7.3×10^{3}
			298.2	− 15.66	−24.64	8.98	0.03
			500	− 14.56	−39.95	25.39	2×10^{-3}
$HCl + HCl \rightleftharpoons HCl \ldots HCl$[f]	− 8.84	− 4.30	100	− 5.81	− 7.19	1.38	0.19
	(−8.95)[g]		298.2	− 5.46	−21.21	15.75	1.7×10^{-3}
			500	− 4.12	−33.88	29.75	7.8×10^{-4}
$Ar + ClF \rightleftharpoons Ar \ldots ClF$	− 2.699	− 2.18	100	− 2.64	− 6.63	3.98	8.3×10^{-3}
			298.2	− 2.68	−19.84	17.16	9.9×10^{-4}
			500	− 2.69	−33.28	30.59	6.4×10^{-4}

[a] Taken from Ref.[166]), [b] Taken from Ref.[85]), [c] Taken from Ref.[167]), [d] Taken from Ref.[168]),
[e] Taken from Ref.[38]), [f] Taken from Ref.[40]), [g] Taken from Ref.[84])

formed using ab initio molecular constants. That the dimer exists in two isomeric forms was taken into consideration.

The importance of the TΔS term warrants further attention. Only for $(LiH)_2$ is the TΔS term considerable smaller than ΔH^0_T. With all other complexes (Table 8) the TΔS terms are either comparable with ΔH^0_T or larger. The HF ... HCl, HCl ... HCl, HF ... ClF and Ar ... ClF complexes are thus formed in processes which are entropy controlled. The values of ΔH^0_{298} and $T\Delta S^0_{298}$ for the complexes given in Table 8 (except for $(LiH)_2$) are worth noting. While the former are between —49.3 and —2.7 kJ/mol, the latter lie in a narrower interval (—32.6, —19.8 kJ/mol). Obviously, the TΔS term does not depend on the complex type as markedly as the enthalpy term.

As mentioned above, the thermodynamic data given in Table 8 are based on ab initio SCF calculations with a medium basis set. For some complexes better quality molecular characteristics (optimum geometry, vibrational wavenumbers, ΔE) are available; comparison of the respective thermodynamic characteristics enables us to test the quality of relatively easily accessible values based on medium basis sets. The thermodynamic characteristics of HF ... HF formation based on SCF calculations with an extended basis set[36,79] (F: [7s4p2d], H: [4s1p]) are given in Table 9. Comparison of the calculated characteristics with those in Table 8 indicates that: a) 4-31G ΔH^0_T differs considerably from the value obtained using the extended basis set; this is due both, to the difference in ΔE as well as to that in the zero-point vibrational energy; b) TΔS is not as sensitive to the basis set used. A complete set of input molecular data is not available for the other complexes. With the $(H_2O)_2$ dimer, the changes in thermodynamic characteristics derived from ΔE and from the intersystem vibrational frequencies with different accuracy[91] can be examined. The thermodynamic characteristics of the formation of $(H_2O)_2$ are given in Table 10. ΔE and the intersystem vibrational frequencies are given by a) ab initio SCF (4-31G) calculation[54], b) H.F. calculations[170], c) H.F. plus dispersion energy calculation[171] and d) the empirical potential[172]. All the other characteristics were taken from the ab initio SCF 4-31G calculation[54]. The results in Table 10 again indicate that the ΔG^0 values depend largely on the accuracy of ΔH^0. The intermolecular vibration frequencies play a crucial role in the evaluation of vibrational partition functions. Table 11 gives the thermodynamic values for H_2O ... HF and HCN ... HF complex formation, using experimental intersystem frequencies; all the other molecular characteristics are based on ab initio calculations with 6-31G[41] and 4-31G[43] basis sets. Comparison of the

Table 9. Thermodynamic characteristics for the formation (kJ/mol) of the HF ... HF dimer; input data are based on ab initio SCF calculation with an extended basis set[36,79]

ΔE	ΔH^0_0	T(K)	ΔH^0	$T\Delta S^0$	ΔG^0	K
—15.91	—10.42	100	—12.10	— 7.28	—4.82	329.6
		298.2	—12.29	—22.37	10.08	0.017
		500	—11,24	—36.21	24.97	0.002

Table 10. Thermodynamic characteristics for the formation (kJ/mol) of the $(H_2O)_2$ dimer at 298.2 K; ΔE and intersystem vibrations taken from Ref.[91], intrasystem vibrations determined by ab initio SCF calculation[54] (taken from Ref.[165])

Method of determining ΔE and intersystem vibrations	ΔE	ΔH_0^0	ΔH^0	$T\Delta S^0$	ΔG^0	K
Ab initio SCF (4-31G)	−34.39	−24.28	−26.06	−24.40	−1.66	1.95
Ab initio H.F.	−20.46	−11.10	−12.51	−23.88	11.37	0.0102
Ab initio H.F. plus dispersion energy	−27.11	−17.17	−18.50	−21.46	2.96	0.3030
Empirical potential	−24.10	−13.86	−15.27	−22.40	7.18	0.0564

Table 11. Thermodynamic characteristics for the formation (kJ/mol) of H_2O ... HF and HCN ... HF (298.2 K); intermolecular frequencies taken from experiment[a], all other input data based on ab initio SCF calculations with 6-31G[41] and 4-31G basis sets[43] (cf. Table 8)

Complex	ΔE	ΔH_0^0	ΔH^0	$T\Delta S^0$	ΔG^0	K
H_2O ... HF	−56.47	−48.44	−51.37	−28.36	−23.01	1.07×10^4
HCN ... HF	−38.38	−30.43	−32.23	−28.44	− 3.79	4.62

[a] Ref.[42,87]

corresponding entries of Tables 8 and 11 indicates that ΔH^0 and ΔH_T^0 differ only slightly; the opposite is true, however, for TΔS.

In conclusion, it can be stated that accurate determination of ΔH_T^0 requires accurate ΔE values; the accuracy of TΔS is affected primarily by the intermolecular frequency values. This conclusion is especially important for larger complexes, for which the evaluation of the optimum geometry and the determination of all the force constants is very complicated. In the calculation of thermodynamic characteristies the following procedure might be useful:

i) the subsystem geometry is either obtained experimentally or optimized and the force constants are evaluated;
ii) the intersystem coordinates are optimized and, again, the force constants are calculated;
iii) the vibrational frequencies of the complex are determined using the force constants of subsystems and intermolecular force constants.

The thermodynamic treatment discussed so far employed the harmonic approximation. Its use for weak intermolecular interactions is, however, not entirely justified. With strong vdW molecules (of course, except strong hydrogen-bonded complexes, vide supra) we can hope that the anharmonicity (in connection with thermodynamic treatment) does not play a crucial role. There exists another complication concerning complexes possessing a double-minimum on energy hypersurface. If the minima are separated by low energy barrier the harmonic approximation is not adequate[63,173]. The role of anharmonicity is essential with true vdW molecules. Unfortunately, except for the H_2O ... HF and $(H_2O)_2$ complexes no complete experimental thermodynamic characteristics are available for the formation of those vdW complexes which have been treated theoretically (see Table 8). Comparison of the theoretical characteristics obtained by the harmonic approach with the experimental characteristics (including anharmonicity) could help understand the role played by anharmonicity; good agreement was found between calculated and experimental $T\Delta S^0$ for the studied complexes. The only example for which the role of anharmonicity can be elucidated is the Xe ... Xe vdW molecule [174]. The vibrational energy levels for Xe ... Xe were obtained by the WKB method using the Lennard-Jones (6–12) potential (the respective parameters were found from experimental measurements). The thermodynamic characteristics of Xe ... Xe were calculated using the vibrational partition function. The values of $(G-H_0)$, $(H-H_0)$ and TS (T = 300 K) evaluated using both the anharmonic and harmonic approach are given in parentheses; all the values are given in kJ/mol: (−1.21, 5.98, 7.19; −0.83, 4.55, 5.38). The differences are, of course, significant, but not as large as might be expected. The following argument for the role of anharmonicity in vdW molecules is only indirect and not really strong. When plotting ΔH^0_{298} against $T\Delta S^0_{298}$ for various hydrogen-bonded complexes and true vdW molecules, a rather close linear relationship was obtained (vide infra). The set under study included complexes for which ΔH and TΔS were evaluated theoretically using the rigid rotor — harmonic oscillator approximation as well as complexes for which the enthalpy and entropy changes were determined experimentally. The first class included the following complexes (the value of ΔH^0_{298} in kJ/mol is given in parentheses, cf. Table 8): H_2O ... HF (−49), HCN ... HF (−32), HF ... HF (−28), H_2O ... H_2O (−26), HF ... HCl (−16), HCl ... HCl (−9), Ar ... ClF (−2). The second group contained the following complexes: CF_3CH_2OH ... CF_3CH_2OH (−20)[175], CH_3OH ... CH_3OH (−13)[176], NO^+ ... N_2 (−5)[177]. Because the linear relationship is satisfied by complexes for which ΔH and TΔS were evaluated theoretically (the harmonic approach), as well as by complexes for which ΔH and TΔS were determined experimentally (anharmonicity included), it cannot be expected that anharmonicity will play a very pronounced role with the type of vdW molecules.

It follows from the preceding discussion that the equilibrium constant for complex formation evaluated using the rigid rotor-harmonic oscillator approximation, with molecular constants derived from ab initio SCF calculations with a medium basis set (of DZ quality), is not very accurate. Comparison of the ΔG^0 values calculated using extended and medium basis sets indicates that the major uncertainty in ΔG^0 is derived from ΔH^0. $T\Delta S^0$ is not as dependent on the basis set used. Furthermore, it is evident that the entropy term plays an extremely important role in complex formation; neglecting it may result not only in quantitative, but even in qualitative failure.

With strong interactions (i.e. formation of common molecules) there is a linear relationship between ΔH and TΔS in series of structurally related systems. It would be interesting to discover whether a similar relationship also exists for weak intermolecular interactions. In this connection, over one hundred vdW complexes, for which ΔH and TΔS are known either from experiment or from theoretical calculations have been investigated. The results are presented in Fig. 4. It is obvious that the set of points can be split into three subsets. The first (individual points designed by +) forms a broad band; this subset includes $X^\pm$. mA complexes

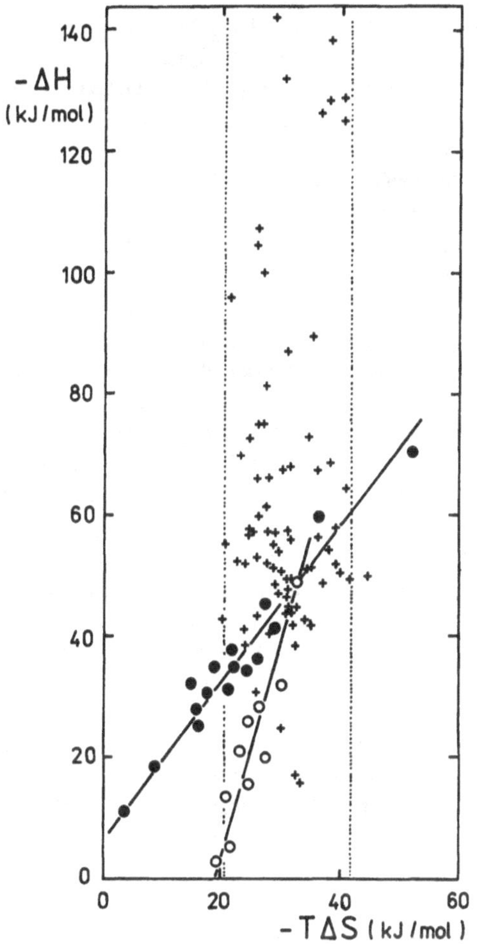

Fig. 4. ΔH Plotted against TΔS for Charge-Transfer Complexes (●), Hydrogen-Bonded Complexes (Including Ar ... ClF) (○) and Hydrated and Solvated Ions (+)

($X^{\pm}$ is a cation or an anion, A is a molecule, m = 1–8), for which ΔH and TΔS were determined experimentally ($Li^+ \cdot mH_2O$, $Na^+ \cdot mH_2O$, $K^+ \cdot mH_2O$, $Rb^+ \times mH_2O$, $Cs^+ \cdot mH_2O$ [178]; $H^+ \cdot mHCN$, $H^+ \cdot mCH_3CN$, i—$C_3H_7^+ \cdot HCN$, t—$C_4H_9^+ \cdot HCN$ [179]; $H^+ \cdot mH_2O$, $H^+ \cdot mCH_3OH$, $H^+ \cdot mCH_3OCH_3$ [180]; $NH_4^+ \times mNH_3$, $NH_4^+ \cdot mH_2O$ [181]; $OH^- \cdot mH_2O$, $F^- \cdot mH_2O$, $Cl^- \cdot mH_2O$, $Br^- \cdot mH_2O$, $I^- \cdot mH_2O$, $CN^- \cdot mH_2O$, $NO_2^- \cdot mH_2O$, $NO_3^- \cdot mH_2O$ [182]; $CH_5^+ \cdot mCH_4$ [183]). Within this class of complexes ΔH^0 varies within much broader limits than $T\Delta S^0$. The average value of the $T\Delta S^0$ terms equals —30.8 kJ/mol. The second subset (points designed by •, Fig. 4) consists of charge-transfer complexes, for which ΔH and TΔS were also determined experimentally. (TCNE ... p-xylene, o-xylene, mesitylene, durene; I_2 ... dimethylsulfide, diethylsulfide, tetrahydrothiophene[184]; TCNE ... benzene, toluene, naphthalene[185]; $CO(CN)_2$ ... furan, thiophene, tetrahydrothiophene, tetrahydrofuran, diethylether, diethylsulfide[186]). Finally, the third subset (points designed by ○, Fig. 4) consist of seven hydrogen-bonded complexes and true vdW molecules for which ΔH and TΔS were evaluated theoretically (cf. Table 8) and three complexes ($CF_3CH_2OH ... CF_3CH_2OH$ [175]), $CH_3OH ... CH_3OH$ [176]), $NO^+ ... N_2$ [177]) for which the values were found experimentally. Splitting of the total set of ΔH and TΔS values into three subsets could be tentatively explained on the basis of the structural features of the complexes. Specifically, while the studied charge-transfer complexes are frequently of the sandwich type, the hydrogen-bonded complexes and true vdW molecules are of the contact type (formation of a quasilinear vdW system). It is evident that the relationships given above are useful for interpolation purposes, e.g., for estimation of entropy changes from calculated enthalpy changes. Its physical significance should not, however, be overestimated. Specifically, Exner[187]) and Krug et al.[188]) discuss the danger of inferring too much from these relationships, as most reported examples of ΔH—ΔS compensation appear to be statistical compensation patterns, because both parameters are generally derived from the same set of data[189]). Exner presented[187]) convincing arguments why regression analysis cannot be used under these circumstances. Various aspects of relationships between ΔH^0 and ΔS^0 were investigated by Tomlinson et al.[190]) in connection with biological activity studies (the interaction of phenothiazines with alkyl sulphates and the interactions between large organic ions).

The thermodynamic characteristics of monohydration of monoatomic cations are listed in Table 12. The ΔE^{SCF} values for all the complexes were obtained using extended basis sets; for details see Ref.[78, 79]). For the majority of the complexes studied, the calculated thermodynamic values, based on the rigid rotor-harmonic oscillator approximation, can be compared with the corresponding experimental characteristics[178, 191]).

The agreement between the calculated and experimental entropy changes for the $Li^+ ... OH_2$ complex is worth noting. For $Na^+ ... OH_2$, $K^+ ... OH_2$ and $Cl^- ... HOH$, the difference between calculated and experimental TΔS values is about 2 kJ/mol. Only for the $F^- ... HOH$ complex does the theoretical value differ significantly from

Table 12. Thermodynamic characteristics for the formation (kJ/mol) of various ion-water complexes. Experimental data (Ref.[178, 191])) are given in parentheses

Complex	ΔE^{SCF} [a]	ΔH^0_0	ΔH^0_{298}	$T\Delta S^0_{298}$	ΔG^0_{298}	log K_{298}
$Li^+ ... OH_2$ [b]	−147.3	−133.5	−137.9	−28.4	−109.5	19.2
			(−142.3)[c]	(−28.7)[c]	(−106.7)[c]	
$Li^+ ... OH_2$ [d]	−147.3	−138.8	−143.2	−28.4	−114.8	20.1
			(−142.3)[c]	(−18.7)[c]	(−106.7)[c]	
$Na^+ ... OH_2$ [d]	−100.4	− 94.1	− 97.5	−26.8	− 70.7	12.4
			(−100.4)	(−28.6)	(− 73.9)	
$K^+ ... OH_2$ [d]	− 69.5	− 64.05	− 67.1	−25.3	− 41.8	7.33
			(− 74.9)	(−27.4)	(− 47.7)	
$Mg^{2+} ... OH_2$ [b]	−325.5	−314.8	−314.5	−30.4	−284.1	49.8
$Al^{3+} ... OH_2$ [b]	−753.1	−746.1	−746.2	−31.3	−714.9	125.2

[a] ΔE^{SCF} and force constants (extended basis set) for Li^+, Na^+, K^+ and for the Mg^{2+}, Al^{3+} complexes from Ref.[78 and 79]), respectively

[b] All frequencies taken from the energy hypersurface

[c] Obtained by extrapolation

[d] OH stretch and HOH bend taken from isolated water

the experimental value. This is understandable because of failure of the harmonic approximation with the anion complexes of this type (cf. comment in Sect. 2.3). Therefore, the above mentioned result for Cl^- ... HOH is fortuitous.

The ΔH^0_{298} values warrant further comment. The enthalpy changes were calculated at various levels of sophistication. At the first level (termed A) all the vibrational frequencies of the complex were taken from the potential energy hypersurfaces, while, at the second level (termed B), force constants corresponding to OH stretching and HOH bending were taken from the isolated water molecule (experimental values[80]). Experimental vibrational frequencies of isolated water were used[80]. Level A should be preferred from the theoretical viewpoint. The Li^+ ... OH_2 complex is the only one for which the vibrational frequencies were evaluated at both levels. Table 12 indicates that ΔH^0_{298} (level A) is smaller than ΔH^0_{298} (level B) and also that the experimental ΔH^0_{298} value is closer to the theoretical ΔH^0_{298} value (level B). It must be taken into account, however, that the theoretical ΔH^0_{298} value is based on ΔE^{SCF} alone. ΔE^{COR} values for Li^+ ... OH_2 were calculated by the semiempirical Wigner method[78] and by the CI procedure[192]. The entries in Table 13 are based on more reliable CI results. As ΔE^{COR} is positive for Li^+ ... OH_2, the difference between the theoretical (level A) and experimental ΔH^0_{298} values is more pronounced. Considering the reliability of the basis set used it seems possible that the experimental value (which is, in fact, extrapolated) could be associated with some error. With other complexes, for which the ΔE^{SCF} value is considerably smaller than that for Li^+ ... OH_2, a smaller difference between the vibration frequencies evaluated at levels A and B can be expected; hence, the difference between the ΔH^0_{298} value calculated at both levels should also be smaller (compared with Li^+ ... OH_2). Thus, considering that ΔE^{COR} was neglected, the agreement between the experimental and theoretical ΔH^0_{298} values for the other complexes listed in Table 12 is satisfactory. For Mg^{2+} ... OH_2 and Al^{3+} ... OH_2 (for which experimental results were not avail-

Table 13. Thermodynamic characteristics for the formation (kJ/mol) of the Li^+ ... OH_2 und F^- ... HOH complexes. Experimental data are given in parentheses (Refs.[178, 191])

Complex	ΔE^{SCF} [a]	$\Delta E^{SCF} + \Delta E^{COR}$ [b]	ΔH^0_{298}
Li^+ ... OH_2 [c]	−147.3	−142.4	−133.0 (−142.3)
Li^+ ... OH_2 [d]	−147.3	−142.4	−138.3 (−142.3)
Li^+ ... OH_2 [c]	−151.0[b]	−146.1	−136.7 (−142.3)
F^- ... HOH[d]	− 99.2	−107.6	− 98.9[e] (− 97.5)

[a] Taken from Ref.[78], extended basis set
[b] Taken from Ref.[192], extended basis set
[c] All frequencies taken from the energy hypersurface
[d] OH stretch and HOH bend taken from isolated water
[e] This value is of limited significance because of doubtful quality of vibrational frequencies (see Sect. 2.3)

able), considering that ΔE^{COR} was neglected, the theoretical ΔH^0_{298} values (evaluated at level A) represent the upper limit for ΔH^0_{298}, i.e. the experimental values can be expected to be slightly larger in absolute values.

Finally the values of ΔH^0_0 and ΔH^0_{298} should be considered. The difference between ΔE and ΔH^0_0 constitutes the zero-point vibrational energy (ΔH^0_{VIB}). Its values for the complexes studied are given in the first column of Table 14. It can be seen that ΔH^0_{VIB} is rather sensitive to the calculation method (in the sense of levels A and B) and decreases for complexes containing heavier isovalent ions. The thermal changes in the enthalpy from absolute zero to room temperature (termed $\Delta H'_{298}$) are listed in Table 14, column 2. This contribution is sometimes felt to be negligible; evidently, this is not so. With complexes containing heavier ions, this term almost compensates the zero-point vibrational energy. The following columns in Table 14 give the contributions from translational, rotational and vibrational partition functions. The rotational contributions are almost zero, the translational contributions are constant for all the complexes investigated and are negative. Finally, the positive vibrational contributions become more important with complexes containing heavier isovalent ions. The last column of Table 14 gives the sum of the zero-point vibrational energy and of the thermal changes in ΔH^0 (adding this sum to ΔE gives ΔH^0_{298}). These values, which are smaller than ΔH^0_{VIB}, could be compensated by ΔE^{COR} (except for $Li^+ \ldots OH_2$ where ΔE^{COR} is positive). This assumption was confirmed for $F^- \ldots HOH$ (see Table 13). With heavier ion ... water complexes, where ΔE^{COR} attains more negative values and $\Delta H^0_{VIB} + \Delta H'_{298}$ has less positive values, ΔH^0_{298} can be expected to be more negative than ΔE^{SCF}.

In conclusion, it should be noted that the ΔH^0_{298} values given in Table 13 represent the most accurate theoretical data for the $Li^+ \ldots OH_2$ and $F^- \ldots HOH$ complexes. Analogous data given in Ref.[78] are based on less reliable ΔE^{COR} values, those from Ref.[192] neglect the thermal change in the enthalpy.

Now, the statistical thermodynamic treatment of equilibria of true vdW molecules will be mentioned. It has been shown[193] that there are considerable differences be-

Table 14. Zero-point vibrational energies (ΔH^0_{VIB}) and temperature dependent terms of ΔH (their sum is termed $\Delta H'_{298}$) calculated for various ion ... water complexes (298 K). Energies in kJ/mol. For references on force constants see Table 12

Complex	ΔH^0_{VIB}	$\Delta H'_{298}$	ΔH^{TRAN}	ΔH^{ROT}	ΔH^{VIB}	$\Delta H^0_{VIB} + \Delta H'_{298}$
$Li^+ \ldots OH_2$ [a]	13.8	−4.37	−6.20	0.04	1.79	9.4
$Li^+ \ldots OH_2$ [b]	8.5	−4.37	−6.20	0.04	1.79	4.1
$Na^+ \ldots OH_2$ [b]	6.3	−3.46	−6.20	0.04	2.70	2.9
$K^+ \ldots OH_2$ [b]	5.5	−3.06	−6.20	0.04	3.10	2.4
$Mg^{2+} \ldots OH_2$ [a]	10.7	−5.16	−6.20	0.04	1.00	5.5
$Al^{3+} \ldots OH_2$ [a]	7.0	−5.51	−6.20	0.04	0.65	1.5
$F^- \ldots HOH$ [b,c]	13.3	−4.71	−6.20	0.04	1.45	8.7
$Cl^- \ldots HOH$ [b,c]	6.6	−3.19	−6.20	0.04	2.97	3.4

[a] All frequencies taken from energy hypersurface
[b] OH stretch and HOH bend taken from isolated water
[c] See note [e] in Table 13

tween the standard treatment and that adequate for true vdW molecules. First, the exponential term mostly approaches unity and therefore K is not an exponential function of the temperature. Second, to a first approximation all the levels are equally occupied at normal experimental temperatures. Third, the usual factorization of the internal partition function into vibrational and rotational contributions is not permitted because of strong interaction between the vibrational and rotational motions. The weak binding of components of a true vdW molecule is responsible for the significant influence of the rotational motion on the shape of the potential energy curve.

Mahan and Lapp[193] dealt with interaction between alkali (A) and noble gas (N) atoms (the original[193] symbols are retained):

$$A + N \rightleftharpoons A \ldots N .$$

The equilibrium constant, K, assumes the form:

$$K = \frac{n_{AN}}{n_A n_N} = \frac{Q_{AN}}{Q_A Q_N} e^{\varepsilon_0/kT}, \tag{18}$$

where n_i is the number of i species, Q is the total partition function and ε_0 is the maximum well depth. Q is separated into translational (Q′) and internal contributions (Q*):

$$Q = Q'Q^* , \tag{19}$$

$$Q'_A = \left(\frac{2\pi m_A kT}{h^2}\right)^{3/2} V . \tag{20}$$

As ε_0 assumes values of about 10–40 cm^{-1} then $\varepsilon_0 \ll kT$ or at least $\varepsilon_0 < kT$ for the temperature region considered and therefore the exponential term can be omitted:

$$K \approx \left(\frac{h^2}{2\pi\mu kT}\right)^{3/2} \frac{Q^*_{AN}}{Q^*_A Q^*_N}, \tag{21}$$

where μ is the reduced mass of A ... N. With some additional plausible approximations the following expression was obtained for the n_{AN}/n_A ratio:

$$\frac{n_{AN}}{n_A} \approx \frac{p}{kT}\left(\frac{h^2}{2\pi\mu kT}\right)^{3/2} \sum_{l=0}^{l_{max}} (2l+1)\,\Omega_l, \tag{22}$$

where p is the total system pressure, l is the rotational angular momemtum quantum number and Ω_l is the number of vibrational levels. This ratio (22) has been tabulated for nearly all possible pairs between Ne, Ar, Kr, Xe and Li, Na, K, Rb, Cs.

Another important procedure for calculating the mole fraction of dimers using the Lennard-Jones potential was suggested by Stogryn and Hirschfelder[194]. Equations are presented which permit calculation of three components of the second

Table 15. Experimentally determined equilibrium constants (atm^{-1}) for formation of various vdW molecules[a]

Complex	T(K)	K	References
H_2 ... Ar	86	44[b]	196
O_2 ... O_2	71	0.003	197
Br_2 ... Br_2	408	0.013	198
I_2 ... I_2	605	0.008	118
NO ... NO	121	129	82
NO^+ ... N_2	178	152	177
HF ... H_2O	315	0.24	85
HF ... $(CH_3)_2O$	303	14	88
HF ... $CH_3C_2H_5O$	303	20	88
HF ... $(C_2H_5)_2O$	303	10	88
H_2O ... H_2O	376	0.011	167
NO_2 ... NO_2	298	6.89	52
CH_3OH ... CH_3OH	337	0.030	176
CH_3OH ... CH_3NH_2	313	0.34	199
CH_3OH ... $(CH_3)_2NH$	313	0.38	199
CH_3OH ... $(CH_3)_3N$	313	0.49	199
CH_3OH ... $(C_2H_5)_3N$	313	0.51	199
CF_3CH_2OH ... CF_3CH_2OH	338	0.095	175
CF_3CH_2OH ... $(CH_3)_3N$	309	6.2	200
$(CH_3)_2CO$ ... $(CH_3)_2CO$	341	0.059	201

[a] $A + B \rightleftharpoons A \dots B; K = [A \dots B]/([A][B])$
[b] cm^3/mole

virial coefficient which are related to the equilibrium constant for the formation of bound double molecules, of metastable double molecules and for interacting molecules which separate after the interaction. In addition to the rare gas atoms, O_2, CO, CO_2, CH_4, C_2H_4 and C_2H_6 were also studied.

Specific aspects of equilibrium constants for formation of vibrationally excited vdW molecules were analyzed by Ewing[195].

This section can be closed by listing numerical values of experimental equilibrium constants for the formation of vdW molecules (Table 15). Values are given only for selected temperatures; values for other temperatures may be found in the references listed in Table 15. Numerous equilibrium constants for charge-transfer complex formation can be found in Refs.[184–186].

4 Reactivity

Reactions involving vdW systems can be divided into two groups. The first comprises formation, transformation and decomposition of vdW molecules. Participation of these molecules in common chemical reactions constitutes the second group.

4.1 Rate of Formation, Transformation and Decomposition of vdW Molecules

It is possible to distinguish between processes involving rather strong vdW complexes and those in which true vdW molecules participate. The former group does not require special treatment: the standard statistical-thermodynamic approach to equilibria and rates (Eyring's ART) is applicable in a rather straightforward way. The rate of transformation between equivalent forms of HX ... HX (here X = F, Cl) belongs to this class[40,65]. Quantum mechanical tunnelling should be included in this and various related processes.

Kinetic analysis of the first step in the formation of the Ar ... Ar dimer from the pure monomer was carried out[202]. The following two-step mechanism was considered:

$$A + A \underset{k_2}{\overset{k_1}{\rightleftharpoons}} A_2^*, \tag{23}$$

$$A_2^* + A \xrightarrow{k_3} A_2 + A. \tag{24}$$

It was found, however, that the specific features of the free-jet nucleation of argon require another scheme:

$$Ar + Ar + Ar \underset{k_r}{\overset{k_f}{\rightleftharpoons}} Ar \ldots Ar + Ar. \tag{25}$$

Several forms of the rate law were investigated and best results were obtained for the inverse power expression, $k = BT^{-n}$, where n = 2 or 3.

Increasing theoretical and experimental attention has been paid to the vibrational relaxation and predissociation of vibrationally excited vdW molecules[58,195,203-205]. Efficient vibrational relaxation[195] takes place through the collisional mechanism[195]:

$$XZ^* + Y \rightarrow XZ + Y, \tag{26}$$

where XZ^* is the vibrationally excited molecule. At low temperatures in the presence of a third body (which removes the kinetic energy), the vdW molecule $XZ^* \ldots Y$ may be formed. If its vibrational energy is greater than the dissociation energy of the vdW bond, vibrational predissociation may occur. In the cited paper by Ewing[195] the equilibrium constants for formation of vibrationally excited vdW molecules, their lifetimes and the vibrational relaxation rate constants were considered. The lifetimes of vdW molecules $H_2^* \ldots H_2$, $N_2^* \ldots N_2$, $O_2^* \ldots O_2$, $HF^* \ldots HF$, $HCl^* \ldots HCl$ and $NO^* \ldots NO$ are between 10^4 and 10^{-11} s.

4.2 Participation of van der Waals Molecules in Common Chemical Reactions

When investigating the fate of reactants passing gradually, via metastable intermediates (minima on energy hypersurfaces) and activated complexes (saddle points on energy hypersurfaces), to products through mono, bi and termolecular reactions, a few possibilities where vdW systems might participate appear. Both metastable intermediates and activated complexes possess some characteristic features of vdW systems (e.g., shallow, anharmonic potential wells, unusual lengths of some bonds).

There is another possibility (or even necessity) for participation of vdW systems in bimolecular reactions: at the foot of the energy barrier separating reactants from products are minima corresponding to vdW complexes formed between the reaction components on both sides of the energy barrier. Let us illustrate this on a reaction which was carefully studied by experimental and theoretical methods (for the original papers see Ref.[206]):

$$F + H_2 \rightarrow HF + H \,. \tag{27}$$

Calculations on the reactants and products, on the activated complex[207] and on the vdW molecules[208, 209] permit tentative drawing of a reaction profile (Fig. 5) and drawing up of the following reaction scheme:

$$\underset{}{F + H{-}H} \rightleftharpoons \underset{\text{vdW I}}{F \ldots H{-}H} \rightleftharpoons [F \ldots H \ldots H]^{\neq} \rightleftharpoons \underset{\text{vdW II}}{F{-}H \ldots H} \rightleftharpoons F{-}H + H \,. \tag{28}$$

It might be supposed that such shallow minima as those typical for the majority of vdW molecules cannot significantly influence the course of chemical reactions. It seems that this is actually true for numerous reactions. But the evidence presented in the next paragraph shows convincingly that vdW intermediates can sometimes influence the reaction course dramatically.

It seems that vdW systems might assume a particularly important role in

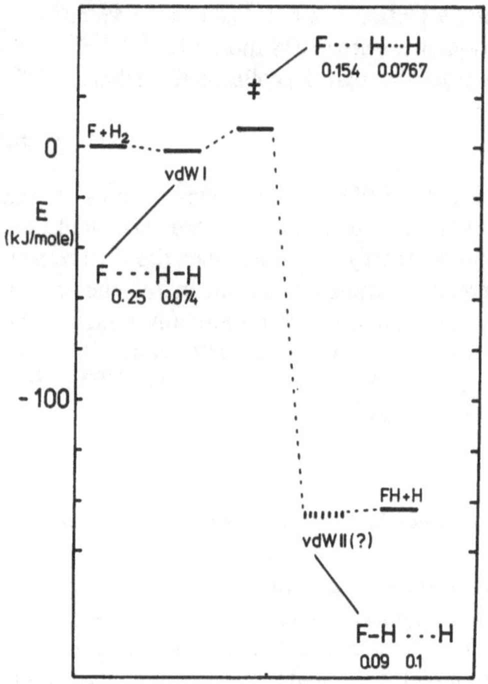

Fig. 5. Estimated Reaction Profile of the Process (28). ≠ indicate an activated complex, bond lengths are given in nm

termolecular reactions where two mechanisms are considered plausible[210]. The first is based on simultaneous interaction of A, B and C and leads to the formation of an activated complex which then decompose to give the products:

$$A + B + C \rightarrow [A \ldots B \ldots C]^{\neq} \rightarrow \text{products}\,. \qquad (29)$$

An alternative mechanism which was felt[210] to be more useful includes a key role for vdW systems. Specifically, a termolecular reaction is considered as the overall result of two successive bimolecular processes. In the first step, a vdW system is formed (in general between any pair of reactants, e.g., A ... B, B ... C, A ... C) and in the second step the vdW system reacts with the remaining component and forms the activated complex which decomposes to give the products.

Recent detailed studies of chemical reactivity, based on utilization of molecular beams, suggest that the second mechanism is more plausible. Herschbach and co-workers[211-213] studied experimental substitution reactions with clusters of $(Cl_2)_n$. No reaction attributable to a bimolecular process was observed between Br_2 and Cl_2 in crossed molecular beams (a collision energy of up to 105 kJ/mol). However, the following reaction proceeds readily, even at thermal collision energies (about 12 kJ/mol):

$$Br_2 + Cl_2 \ldots Cl_2 \rightarrow 2\,BrCl + Cl_2\,. \qquad (30)$$

The energy of the vdW bond is about 4–8 kJ/mol, and the Cl ... Cl bond length amounts to 0.43 nm. A six-centre bond exchange is assumed to occur through a cyclic transition state that dissociates through the open chain structure:

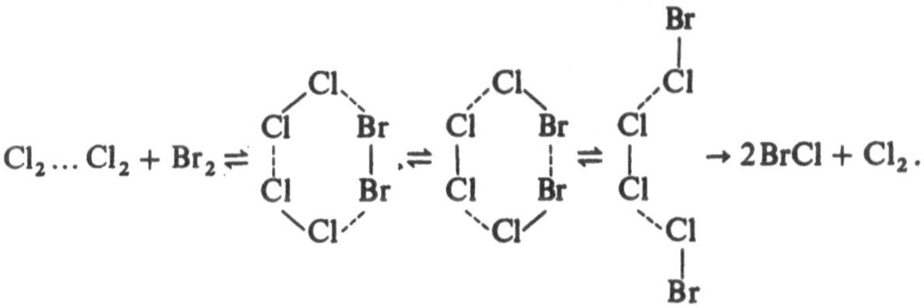

(31)

Note Added in Proof

The geometry of the H_2O ... HF complex was partially optimized using triple zeta (TZ) and TZ + P basis sets[214]. The respective optimum O ... F distances are 0.267 and 0.269 nm. Quadratic and cubic force constants were obtained for both basis sets. On passing from the first to the second basis set, the intermolecular force constant, K_{RR}, decreased by 57%. When passing from the 6-31G basis set to the TZ + P basis set the complex departs from the C_{2v} symmetry.

More accurate ΔH^0 and $T\Delta S^0$ values were obtained for water dimerization by means of thermal conductivity measurements[215]. At 373 K the values are −15.02 and −29.01 kJ/mol, respectively.

The approach to vibrational and thermodynamic properties of Ar ... HX vdW molecules, used by Shin[67], was criticized by le Roy[216] who presented a more advanced analysis: special attention was paid to the vibration-rotation spectrum of Ne ... Ne, to the vibrational spacings and thermodynamic properties of rare gas dimers and Ar ... HX vdW molecules.

Monte Carlo method was used in combination with a sophisticated empirical potential for evaluation of thermodynamic characteristics for $(H_2O)_2$. $T\Delta S^0$ amounts (300 K) to −16.60 kJ/mol[217].

Acknowledgement. We greatly appreciate valuable comments of Dr. Małgorzata M. Szczęśniak (Wrocław), Dr. Z. Herman (Prague), Dr. H.-J. Hofmann (Leipzig), and Dr. J. Sauer (Berlin) to the manuscript of this work. We are indebted to Professors L. A. Curtiss (Argonne), W. Klemperer (Cambridge, Mass.), and H. Lischka (Vienna) for having kindly sent us reprints, preprints and unpublished data.

5 References

1. Ewing, G. E.: Canad. J. Phys. *54*, 487 (1976)
2. Hobza, P., Zahradník, R.: Weak Intermolecular Interactions in Chemistry and Biology, Amsterdam and Prague, Elsevier and Academia, 1980
3. Novick, S. E., Janda, K. C., Klemperer, W.: J. Chem. Phys. *65*, 5115 (1976)
4. Liu, B., McLean, A. D.: J. Chem. Phys. *59*, 4557 (1973)
5. Burgmans, A. L. J., Farrar; J. M., Lee, Y. T.: J. Chem. Phys. *64*, 1345 (1976)
6. Stevens, W. J. et al.: J. Chem. Phys. *60*, 2195 (1974)
7. Farrar, J. M. et al.: Chem. Phys. Lett. *19*, 359 (1973)
8. Cohen, J. S., Pack, R. T.: J. Chem. Phys. *61*, 2372 (1974)
9. Parson, J. M., Siska, P. E., Lee, Y. T.: J. Chem. Phys. *56*, 1511 (1972)
10. Barker, J. A. et al.: J. Chem. Phys. *61*, 3081 (1974)
11. Hariharan, P. C., Kutzelnigg, W.: unpublished results, 1976
12. Schafer, R., Gordon, R. G.: J. Chem. Phys. *58*, 5422 (1973)
13. Lloyd, J., Pugh, D.: Chem. Phys. Lett. *39*, 468 (1976)
14. Riehl, J. W. et al.: J. Chem. Phys. *58*, 4571 (1973)
15. Mintmire, J. W., Sabin, J. R.: Int. J. Quantum Chem., QCS *10*, 213 (1976)
16. LeRoy, R. J., Van Kranendonk, J.: J. Chem. Phys. *61*, 4750 (1974)
17. Henderson, G., Ewing, G. E.: Mol. Phys. *27*, 903 (1974)
18. Tully, F. P., Lee, Y. T.: J. Chem. Phys. *57*, 866 (1972)
19. Henderson, G., Ewing, G. E.: J. Chem. Phys. *59*, 2280 (1973)
20. Smalley, R. E., Wharton, L., Levy, D. H.: J. Chem. Phys. *68*, 671 (1978)
21. Novick, S. E. et al.: Can. J. Phys. *53*, 2007 (1975)
22. Blustin, P. H.: Theoret. Chim. Acta *47*, 249 (1978)
23. Miziolek, A. W., Pimentel, G. C.: J. Chem. Phys. *66*, 3840 (1977)
24. Howard, B. J.: Lecture MOLEC II Conf., Brandbjerg, Højschole, Denmark, Aug. 1978
25. Harris, S. J. et al.: J. Chem. Phys. *63*, 881 (1975)
26. Smalley, R. E., Wharton, L., Levy, D. H.: Acc. Chem. Res. *10*, 139 (1977)
27. Balfour, W. J., Douglas, A. E.: Can. J. Phys. *48*, 901 (1970)
28. Balfour, W. J., Whitlock, R. F.: Can. J. Phys. *53*, 472 (1975)
29. Jaszuński, M., Kochanski, E., Siegbahn, P.: Mol. Phys. *33*, 139 (1977)
30. Farrar, J. M., Lee, Y. T.: J. Chem. Phys. *57*, 5492 (1972)
31. Long, C. A., Henderson, G., Ewing, G. E.: Chem. Phys. *2*, 485 (1973)
32. Goodman, J., Brus. L. E.: J. Chem. Phys. *67*, 4398 (1977)
33. Skaarup, S., Skancke, P. N., Boggs, J. E.: J. Amer. Chem. Soc. *98*, 6106 (1976)
34. Dinerman, C. E., Ewing, G. E.: J. Chem. Phys. *53*, 626 (1970)
35. Butz, H. P. et al.: Z. Phys. *247*, 70 (1971)
36. Lischka, H.: Chem. Phys. Lett. *66*, 108 (1979)
37. Dyke, T. R., Howard, B. J., Klemperer, W.: J. Chem. Phys. *56*, 2442 (1972)
38. Hobza, P., Szczęśniak, M. M., Latajka, Z.: Chem. Phys. Lett., in press

39. Janda, K. C. et al.: J. Chem. Phys. *67*, 5162 (1977)
40. Hobza, P., Čársky, P., Zahradník, R.: Coll. Czech. Chem. Commun., *44*, 3458 (1979)
41. Lister, D. G., Palmieri, P.: J. Mol. Struct. *39*, 295 (1977)
42. Bevan, J. W. et al.: J. Chem. Soc., Chem. Commun. *1975*, 341
43. Curtiss, L. A., Pople, J. A.: J. Mol. Spectrosc. *48*, 413 (1973)
44. Legon, A. C., Millen, D. J., Rogers, S. C.: Chem. Phys. Lett. *41*, 137 (1976)
45. Bevan, J. W. et al.: J. Chem. Soc., Chem. Commun. *1975*, 130
46. Shibata, S.: personal communication by Kobayashi, T., 1978
47. Fredin, L., Nelander, B.: J. Amer. Chem. Soc. *96*, 1672 (1974)
48. Fredin, L., Nelander, B.: Mol. Phys. *27*, 885 (1974)
49. Fredin, L., Nelander, B.: J. Mol. Struct. *16*, 217 (1973)
50. Legon, A. C., Millen, D. J., Mjöberg, P. J.: Chem. Phys. Lett. *47*, 589 (1977)
51. Barton, A. E., Chablo, A., Howard, B. J.: Chem. Phys. Lett. *60*, 414 (1979)
52. Novick, S. E., Howard, B. J., Klemperer, W.: J. Chem. Phys. *57*, 5619 (1972)
53. Barnes, A. J., Orville-Thomas, W. J., Szczepaniak, K.: J. Mol. Struct. *45*, 75 (1978)
54. Curtiss, L. A., Pople, J. A.: J. Mol. Spectrosc. *55*, 1 (1975)
55. Fredin, L., Nelander, B., Ribbegård, G.: Chem. Phys. Lett. *36*, 375 (1975)
56. Dyke, T. R., Muenter, J. S.: J. Chem. Phys. *60*, 2929 (1974)
57. Legon, A. C., Millen, D. J., Rogers, S. C.: J. Chem. Soc., Chem. Commun. *1975*, 580
58. Klemperer, W.: Ber. Bunsenges. *78*, 128 (1974)
59. Novick, S. E. et al.: J. Chem. Phys. *59*, 2273 (1973)
60. Harris, S. J. et al.: J. Chem. Phys. *61*, 193 (1974)
61. Novick, S. E. et al.: J. Amer. Chem. Soc. *95*, 8547 (1973)
62. Wilson, E. B., jr., Decius, J. C., Cross, P. C.: Molecular Vibrations, New York, McGraw-Hill, 1955
63. Janoschek, R.: The Hydrogen Bond. I. Theory, p. 165 (P. Schuster, G. Zundel, C. Sandorfy, Eds.), Amsterdam, North-Holland Publ. Comp. 1976
63a. Sandorfy, C.: The Hydrogen Bond. I. Theory, p. 615 (P. Schuster, G. Zundel, C. Sandorfy, Eds.). Amsterdam, North-Holland Publ. Comp., 1976
64. Colthup, N. B., Daly, L. H., Wiberley, S. E.: Introduction to Infrared and Raman Spectroscopy, New York and London, Academic Press, 1964
65. Curtiss, L. A., Pople, J. A.: J. Mol. Spectrosc. *61*, 1 (1976)
66. Messiah, A.: Quantum Mechanics, Vol. 1, Amsterdam, North-Holland Publ. Comp., 1961
67. Shin, H. K.: Chem. Phys. Lett. *49*, 193 (1977)
68. Ewing, G. E.: Acc. Chem. Res. *8*, 185 (1975)
69. Baylis, W. E.: Phys. Rev. *A1*, 990 (1970)
70. Ewing, G. E.: Ann. Rev. Phys. Chem. *23*, 141 (1972)
71. Ewing, G. E.: Angew. Chem. *84*, 570 (1972)
72. Docken, K., Schafer, T. P.: J. Mol. Spectrosc. *46*, 454 (1973)
73. Allin, E. J. et al.: Appl. Opt. *6*, 1597 (1967)
74. Watanabe, A., Welsh, H. L.: Phys. Rev. Lett. *13*, 810 (1964)
75. Kudian, A. K., Welsh, H. L.: Can. J. Phys. *49*, 230 (1971)
76. Rank, D. H., Glickman, W. A., Wiggins, T. A.: J. Chem. Phys. *43*, 1304 (1965) and papers cited therein
77. Long, C. A., Ewing, G. E.: J. Chem. Phys. *58*, 4824 (1973)
78. Kistenmacher, H., Popkie, H., Clementi, E.: J. Chem. Phys. *59*, 5842 (1973)
79. Lischka, H.: unpublished calculations, 1978, (Mg, Al: (10s6p) → [7s5p]; O: (8s4p2d) → [6s4p2d]; H: (4s1p) → [3s1p])
80. Dunning, T. H., Pitzer, R. M., Aung, S.: J. Chem. Phys. *57*, 5044 (1972)
81. Smith, A. L., Keller, W. E., Johnston, H. L.: J. Chem. Phys. *19*, 189 (1951)
82. Billingsley, J., Callear, A. B.: Trans. Faraday Soc. *67*, 589 (1971)
83. Mannik, L., Stryland, J. C., Welsh, H. L.: Can. J. Phys. *49*, 3056 (1971)
84. Rank, D. H. et al.: J. Chem. Phys. *39*, 2673 (1963)
85. Thomas, R. K.: Proc. Roy. Soc. (London) *A344*, 579 (1975)
86. Millen, D. J.: J. Mol. Struct. *45*, 1 (1978)
87. Thomas, R. K.: Proc. Roy. Soc. (London) *A325*, 133 (1971)
88. Thomas, R. K.: Proc. Roy. Soc. (London) *A322*, 137 (1971)

89. Bertie, J. E., Falk, M. V.: Can. J. Chem. *51*, 1713 (1973)
90. Curtiss, L. A.: J. Mol. Struct. *54*, 239 (1979)
91. Huler, E., Zunger, A.: Chem. Phys. *13*, 433 (1976)
92. Jones, W. J., Seel, R. M., Sheppard, N.: Spectrochim. Acta *A25*, 385 (1969)
93. King, C. M., Nixon, E. R.: J. Chem. Phys. *48*, 1685 (1968)
94. Pącansky, J.: J. Phys. Chem. *81*, 2240 (1977)
95. Walsh, B. et al.: J. Mol. Struct. *72*, 44 (1978)
96. Johansson, A., Kollman, P., Rothenberg, S.: Chem. Phys. Lett. *16*, 123 (1972)
97. Pecul, K., Janoschek, R.: Theoret. Chim. Acta *36*, 25 (1974)
98. Fredin, L., Nelander, B., Ribbegård, R.: Chem. Scripta *7*, 11 (1975)
99. Jönsson, B., Karlström, G., Wennerström, H.: Chem. Phys. Lett. *30*, 58 (1975)
100. Fredin, L., Nelander, B.: Chem. Phys. *15*, 473 (1976)
101. Fredin, L., Nelander, B., Ribbegård, G.: Chem. Phys. *12*, 153 (1976)
102. Nelander, B.: J. Mol. Struct. *50*, 223 (1978)
103. Weller, A.: Pure Appl. Chem. *16*, 115 (1968)
104. Beens, H., Weller, A.: Chem. Phys. Lett. *2*, 140 (1968)
105. Davydov, A. S.: Theory of Molecular Excitations, New York and London, McGraw-Hill, 1962
106. Murrell, J. N.: The Theory of the Electronic Spectra of Organic Molecules, Chapter 7. London, Methuen; New York, J. Wiley, 1963
107. Zahradník, R., Hobza, P.: unpublished calculation 1978
108. Del Bene, J., Jaffé, H. H.: J. Chem. Phys. *48*, 1807 (1968)
109. Tanaka, Y., Yoshino, K.: J. Chem. Phys. *50*, 3087 (1969)
110. Tanaka, Y., Yoshino, K.: J. Chem. Phys. *57*, 2964 (1972)
111. Tanaka, Y., Yoshino, K., Freeman, D. E.: J. Chem. Phys. *59*, 564 (1973)
112. Tanaka, Y., Yoshino, K.: J. Chem. Phys. *53*, 2012 (1970)
113. Tanaka, Y., Yoshino, K., Freeman, D. E.: J. Chem. Phys. *59*, 5160 (1973)
114. Freeman, D. E., Yoshino, K., Tanaka, Y.: J. Chem. Phys. *61*, 4880 (1974)
115. Blickensderfer, R. P., Ewing, G. E.: J. Chem. Phys. *51*, 873 (1969)
116. Ogryzlo, E. A., Sanctuary, B. C.: J. Phys. Chem. *69*, 4422 (1965)
117. Tamres, M., Duerksen, W. K., Goodenow, J. M.: J. Phys. Chem. *72*, 966 (1968)
118. Passchier, A. A., Gregory, N. W.: J. Phys. Chem. *72*, 2697 (1968)
119. Arai, S. et al.: J. Chem. Phys. *68*, 4595 (1978)
120. Tanaka, Y., Yoshino, K., Freeman, D. E.: J. Chem. Phys. *62*, 4484 (1975)
121. Dabrowski, I., Herzberg, G.: J. Mol. Spectrosc. *73*, 183 (1978)
122. Turner, D. W. et al.: Molecular Photoelectron Spectroscopy, London, Wiley-Interscience, 1970
123. Siegbahn, K. et al.: ESCA applied to free molecules, Amsterdam, North-Holland Publ. Comp., 1969
124. Čermák, V., Herman, Z.: Coll. Czech. Chem. Commun. *30*, 169 (1965); Miller, W. H.: J. Chem. Phys. *52*, 3563 (1970)
125. Dehmer, P. M., Dehmer, J. L.: J. Chem. Phys. *67*, 1774 (1977)
126. Ng, C. Y. et al.: J. Chem. Phys. *66*, 446 (1977)
127. Huffman, R. E., Katayama, D. H.: J. Chem. Phys. *45*, 138 (1966)
128. Samson, J. A. R., Cairns, R. B.: J. Opt. Soc. Amer. *56*, 1140 (1966)
129. Wedenejew, W. J. et al.: Energien chemischer Bindungen, Ionisationspotentiale und Elektronenaffinitäten, Leipzig, VEB Deutscher Verlag für Grundstoffindustrie, 1971
130. Niehaus, A.: Ber. Bunsenges. Phys. Chem. *77*, 632 (1973)
131. Hotop, H.: Radiat. Res. *59*, 379 (1974)
132. Čermák, V.: J. Electron Spectrosc. Relat. Phenom. *9*, 419 (1976)
133. Čermák, V., Yencha, A. J.: J. Electron Spectrosc. Relat. Phenom. *11*, 67 (1977)
134. Utsunomiya, C., Nagakura, S., Kobayashi, T.: personal communication, 1979
135. Brown, R. S.: Can. J. Chem. *54*, 1929 (1976)
136. Mostad, A. et al.: Chem. Phys. Lett. *23*, 157 (1973)
137. Hobza, P., Čársky, P., Zahradník, R.: Theor. Chim. Acta *53*, 1 (1979)
138. Margenau, H., Murphy, G. M.: Mathematics of Physics and Chemistry, Princeton N. J., D. van Nostrand, 1956
139. Murrell, J. N., Laidler, K. J.: Trans. Faraday Soc. *64*, 371 (1968)

140. Pancíř, J.: Coll. Czech. Chem. Commun. *40*, 1112, 2726 (1975)
141. Fukui, K.: J. Phys. Chem. *74*, 4161 (1970)
142. Pulay, P.: Application of Electronic Structure Theory (H. F. Schaefer III, Ed.) Vol. 4., New York, London, Plenum, 1977
143. McIver, J. W., jr., Komornicki, A.: Chem. Phys. Lett. *10*, 303 (1971)
144. Pancíř, J.: Theor. Chim. Acta *29*, 21 (1973)
145. Murtagh, B. A., Sargent, R. W. H.: Comput. J. *13*, 185 (1970)
146. Pulay, P.: Theor. Chim. Acta *50*, 299 (1979)
147. Weintraub, H. J. R., Hopfinger, A. J.: Int. J. Quantum Chem. *QBS2*, 1975, 203
148. Slanina, Z., Hobza, P., Zahradník, R.: Coll. Czech. Chem. Commun. *39*, 228 (1974)
149. Warshel, A., Karplus, M.: J. Amer. Chem. Soc. *94*, 5612 (1972); *96*, 5677 (1974)
150. Scott, R. A., Scheraga, H. A.: J. Chem. Phys. *44*, 3054 (1966)
151. Ferguson, T. R., Beckel, C. L.: J. Chem. Phys. *59*, 1905 (1973)
152. Gorlov, Y. I., Ukrainsky, I. I., Penkovsky, V. V.: Theor. Chim. Acta *34*, 31 (1974)
153. Yamaguchi, Y., Dewar, M. J. S.: Computors & Chem. *2*, 25 (1978)
154. Pulay, P.: Mol. Phys. *17*, 197 (1969)
155. Pulay, P.: Mol. Phys. *21*, 329 (1971)
156. Meyer, W., Pulay, P.: MOLPRO Description, München and Stuttgart, Germany, 1969
157. Huber, H., Čársky, P., Zahradník, R.: Theor. Chim. Acta *41*, 217 (1976)
158. Schlegel, H. B., Wolfe, S., Bernardi, F.: J. Chem. Phys. *63*, 3632 (1975)
159. Komornicki, A. et al.: Chem. Phys. Lett. *45*, 595 (1977)
160. Poppinger, D.: Chem. Phys. Lett. *34*, 332 (1975); *35*, 550 (1975)
161. Tachibana, A. et al.: Chem. Phys. Lett. *59*, 255 (1978)
162. Kerns, R. C., Allen, L. C.: J. Amer. Chem. Soc. *100*, 6587 (1978)
163. Hobza, P., Pancíř, J., Zahradník, R.: Coll. Czech. Chem. Commun. *45*, 1323 (1980)
164. Thang, N. D. et al.: Coll. Czech. Chem. Commun. *43*, 1366 (1978)
165. Hobza, P., Čársky, P., Zahradník, R.: Int. J. Quantum Chem. *16*, 257 (1979)
166. Kollman, P., Bender, C. F., Rothenberg, S.: J. Amer. Chem. Soc. *94*, 8016 (1972)
167. Curtiss, L. A., Frurip, D. J., Blander, M.: Chem. Phys. Lett. *54*, 575 (1978)
168. Frank, E. U., Meyer, F.: Z. Elektrochem. *63*, 577 (1959)
169. Slanina, Z.: Coll. Czech. Chem. Commun. *43*, 1974 (1978)
170. Kistenmacher, H. et al.: J. Chem. Phys. *60*, 4455 (1974)
171. Lie, G. C., Clementi, E.: J. Chem. Phys. *62*, 2195 (1975)
172. Shipman, L. L., Owicki, J. C., Scheraga, H. A.: J. Phys. Chem. *78*, 2055 (1974)
173. Špirko, V.: personal communication, 1979
174. Shin, H. K.: Chem. Phys. Lett. *47*, 225 (1977)
175. Curtiss, L. A., Frurip, D. J., Blander, M.: J. Amer. Chem. Soc. *100*, 79 (1978)
176. Renner, T. A., Kucera, G. H., Blander, M.: J. Chem. Phys. *66*, 177 (1977)
177. Turner, D. L., Conway, D. C.: J. Chem. Phys. *65*, 3944 (1976)
178. Džidić, I., Kebarle, P.: J. Phys. Chem. *74*, 1466 (1970)
179. Moet-Ner, M.: J. Amer. Chem. Soc. *100*, 4694 (1978)
180. Grimsrud, E. P., Kebarle, P.: J. Amer. Chem. Soc. *95*, 7939 (1973)
181. Payzant, J. D., Cunningham, A. J., Kebarle, P.: Can. J. Chem. *51*, 3242 (1973)
182. Payzant, J. D., Yamdagni, R., Kebarle, P.: Can. J. Chem. *49*, 3308 (1971)
183. Hiraoka, K., Kebarle, P.: J. Amer. Chem. Soc. *97*, 4179 (1975)
184. Kroll, M.: J. Amer. Chem. Soc. *90*, 1097 (1968)
185. Hanazaki, I.: J. Phys. Chem. *76*, 1982 (1972)
186. Fueno, T., Yonezawa, Y.: Bull. Chem. Soc. Japan *45*, 52 (1972)
187. Exner, O.: Progr. Phys. Org. Chem. *10*, 411 (1973)
188. Krug, R. R., Hunter, W. G., Grieger, R. A.: J. Phys. Chem. *80*, 2335 (1976)
189. Tomlinson, E.: personal communication, 1978
190. Tomlinson, E., Davies, S. S., Mukhayer, G. I.: In Solution Chemistry of Surfactants (K. L. Mittal, Ed.). New York, Plenum, 1979
191. Arshadi, M., Yamdagni, R., Kebarle, P.: J. Phys. Chem. *74*, 1475 (1970)
192. Diercksen, G. H. F., Kraemer, W. P., Roos, B. O.: Theor. Chim. Acta *36*, 249 (1975)
193. Mahan, G. D., Lapp, M.: Phys. Rev. *179*, 19 (1969)
194. Stogryn, D. E., Hirschfelder, J. O.: J. Chem. Phys. *31*, 1531 (1959)

195. Ewing, G.: Chem. Phys. *29*, 253 (1978)
196. Weiss, S.: J. Chem. Phys. *67*, 3840 (1977)
197. Blickensderfer, R. P., Ewing, G. E.: J. Chem. Phys. *47*, 331 (1967)
198. Kokovin, G. A.: Zh. Neorg. Khim. *10*, 287 (1965)
199. Millen, D. J., Mines, G. W.: J. Chem. Soc. Faraday II *70*, 693 (1974)
200. Rice, S. A., Wood, J. L.: J. Chem. Soc. Faraday II *69*, 87 (1973)
201. Frurip, D. J., Curtiss, L. A., Blander, M.: J. Phys. Chem. *82*, 2555 (1978)
202. Milne, T. A., Vandegrift, A. E., Greene, F. T.: J. Chem. Phys. *52*, 1552 (1970)
203. Child, M. S.: Faraday Discuss. Chem. Soc. *62*, 307 (1976)
204. Smalley, R. E., Levy, O. H., Wharton, L.: J. Chem. Phys. *64*, 3266 (1976)
205. Beswick, J. A., Jortner, J.: Chem. Phys. Lett. *49*, 13 (1977)
206. Čársky, P., Zahradník, R.: Int. J. Quantum Chem. *16*, 243 (1979)
207. Bender, C. F. et al.: Science *176*, 1412 (1972)
208. Jaffe, R. L., Morokuma, K., George, T. F.: J. Chem. Phys. *63*, 3417 (1975)
209. Hobza, P., Zahradník, R.: calculation for this review
210. Benson, S. W.: The Foundation of Chemical Kinetics, p. 305, New York, Toronto, London, McGraw-Hill, 1960
211. King, D. L., Dixon, D. A., Herschbach, D. R.: J. Amer. Chem. Soc. *96*, 3328 (1974)
212. Dixon, D. A., Herschbach, D. R.: J. Amer. Chem. Soc. *97*, 6268 (1975)
213. Herschbach, D. R.: Pure Appl. Chem. *47*, 61 (1976)
214. Bouteiller, Y., Allavena, M., Leclerq, J. M.: Chem. Phys. Lett. *69*, 521 (1980)
215. Curtiss, L. A., Frurip, D. J., Blander, M.: J. Chem. Phys. *71*, 2703 (1979)

Intermolecular Interactions and Anesthesia: Infrared Spectroscopic Studies

Ginette Trudeau[1], Jean-Max Dumas[2], Paul Dupuis[1], Maurice Guérin[2], and Camille Sandorfy[1]

1 Département de Chimie, Université de Montréal, Montréal Québec, Canada H3C 3V1
2 Laboratoire de Physico-Chimie des Diélectriques, Université de Poitiers, 86022, Poitiers, France

Table of Contents

1 Introduction

All available evidence indicates that general anesthesia is a matter of molecular associations. While anesthetics are metabolized in the organism, anesthetic action itself does not involve chemical reactions or the breaking or formation of covalent or electrovalent bonds. What anesthetic action involves are changes in molecular associations that are vital for the functioning of the nerve cell membrane. We shall call these changes subchemical reactions: a variety of associations by Van der Waals forces, hydrogen bonds and charge transfer might be perturbed and replaced by others as a consequence of anesthetic action.

The main constituents of cell membranes are lipids and proteins. Therefore, we have to study possible associations between these macromolecules and molecules having anesthetic potency. It is instructive, however, at a preliminary stage, to study simpler systems which exhibit some of the basic patterns of association that occur in membranes. The results of such studies will be described and discussed in the present chapter. The method used has been mainly infrared spectroscopy. While other techniques, like Raman and nuclear magnetic resonance spectroscopy turn out to be more powerful tools in many cases, infrared spectra can be highly informative in others, especially when polar groups are involved.

The rational underlying these studies is as follows: The functioning of the nervous system hinges on the permeability of the cell membrane to ions. (Na^+, K^+ and probably others). The permeability depends on the structure of the ion channels determined by the conformations of the macromolecules (lipids and proteins) forming the membrane. In turn, the right conformations are insured, in addition to their chemical structures by a great number of "weak" interactions. It is then logical to assume that any perturbation that interferes with these weak intermolecular interactions might alter the conditions for the transport of ions in and out of the nerve cells whereby impeding the functioning of the nervous system.

It is not the purpose of this chapter to review available knowledge and theories on anesthesia. Excellent recent reviews have been published by Seeman[1], K. W. Miller and Smith[2], Halsey[3], J. C. Miller and K. W. Miller[4], Kaufman[5], and Denson[6]. Among the fundamental papers in which membrane excitation and anesthetic action are linked to altering the functioning of the ion channels and to conformational changes in the membrane macromolecules we cite those by Changeux, Blumenthal, Kasai and Podleski[7], Woodbury, D'Arrigo and Eyring[8] and Urry, Spisni, Khaled, Long and Masotti[9].

2 Intermolecular Interactions

If we seek an understanding of the mechanism of anesthesia at the molecular level, we have to inquire about the nature of intermolecular forces that might play a role in it.

Van der Waals interactions are usually divided into dipole-dipole, dipole-induced dipole and dispersion (London) contributions. When ions are also present, there are, in addition, ion-ion, ion-dipole and ion-induced dipole contributions. At short distances repulsive interactions become important (See for example,[10]).

Two neutral molecules with no dipole moments can only interact through dispersion forces. Dispersion forces are always attractive, whatever the mutual orientation of the molecules. They are independant of temperature. The interaction energy between two isotropic molecules whose centers are separated by a distance r is:

$$U(r) = \frac{3}{2}\frac{h}{r^6}\left(\frac{\nu_1\nu_2}{\nu_1 + \nu_2}\right)\alpha_1\alpha_2, \tag{1}$$

where the α are the polarizabilities of the molecules and the $h\nu$ their isonization potentials. The latter vary relatively little from molecule to molecule while the polarizabilities might vary a great deal. The quadratic dependance on the α is to be noted.

The interaction between two permanent dipoles is, of course, highly dependant on their mutual orientation. If aligned for maximum attraction or repulsion, their interaction energy is:

$$U(r) = \pm\frac{2\mu_A\mu_B}{\varepsilon r^3}, \tag{2}$$

where the μ are the two dipole moments and ε is the permittivity of the medium between them. Lateral alignment would give:

$$U(r) = \pm\frac{\mu_A\mu_B}{\varepsilon r^3}. \tag{3}$$

The negative sign applies to attraction. For random orientation (gas or ideal solution) the average that is relevant is:

$$U_{av} = -\frac{2}{3}\frac{\mu_A^2\mu_B^2}{r^6} \times \frac{1}{kT}, \tag{4}$$

provided $(\mu_A\mu_B/\varepsilon r^3) \ll kT$.

It is very important to remember that the dependence on r^{-6} only applies to random conditions. The interaction between oriented dipoles is much larger.

The interaction energy between two ions or charged groups with charges Z_Ae and Z_Be is:

$$U(r) = \pm\frac{Z_AZ_Be^2}{\varepsilon r^2}, \tag{5}$$

where the negative sign applies to two charges of different signs.

For ion-dipole interaction we have:

$$U(r) = -\frac{Ze\mu}{\varepsilon r^2}, \tag{6}$$

for alignment of maximum attraction and:

$$U(r) = \frac{Ze\mu^2}{3kTr^4}, \tag{7}$$

averaged over all orientations of the dipole.

Ion-ion, ion-dipole and dipole-dipole interactions are often referred to as electrostatic since they do not involve mutual polarization of the molecules or charge transfer between them.

An ion or permanent dipole can induce a dipole moment in an otherwise non-polar molecule. For ion-induced dipole interaction, the energy is:

$$U(r) = \frac{1}{2}\frac{\alpha Z^2e^2}{\varepsilon r^4}. \tag{8}$$

For the case dipole-induced dipole one obtains for maximal attraction:

$$U(r) = -\frac{4\alpha\mu^2}{r^6}, \tag{9}$$

where μ is the dipole moment of the polarizer and α the polarizability of the molecule which becomes polarized. The total average induction energy for a pair of identical polar molecules is:

$$U(r) = -\frac{2\alpha\mu^2}{r^6}. \tag{10}$$

Allowing for mutual polarization between two molecules with dipole moment μ_1 and μ_2:

$$U(r) = -\frac{1}{r^6}(\alpha_1\mu_2^2 + \alpha_2\mu_1^2). \tag{11}$$

To a first approximation it is independent of temperature. It is to be noted that the dipole-induced dipole interaction depends on r^{-6} whether randomized or not. This is connected with the fact that turning the permanent dipole by 180° would not change the sign of the energy in this case for this reverses also the direction of the induced dipole. Only the numerical constant changes as a result of averaging.

The sum of ion-induced dipole and dipole-induced dipole energies are often referred to as induction energy.

Thus for two polarizable dipole molecules in the gas phase all attractive interactions depend on the inverse of the sixth power of their mutual distance. For two identical molecules, neglecting higher terms (quadrupole interactions etc.).

$$\begin{aligned} U(r) &= \frac{b}{r^n} - \frac{1}{r^6}\left[\frac{2}{3}\frac{\mu^4}{kT} + 2\mu^2\alpha + \frac{3}{4}h\nu\alpha^2\right] \\ &= \frac{b}{r^n} - \frac{C}{r^6}. \end{aligned} \tag{12}$$

For further insight the reader might consult the work of Davies[10] on which much of the above summary is based, or Pryde[11] or other treatises.

As a consequence of averaging, the most important contribution to intermolecular energy comes from the dispersion term for all but very polar molecules. Classic examples to illustrate this are given by Davies ([10], p. 166). The dispersion term is preponderant for CO and HCl, the dispersion and electrostatic terms are similar for NH_3 and the electrostatic contribution is the largest for H_2O. The induction term is relatively small in all cases.

Conditions are very different in condensed phases, however. As has been shown above all but the ion-ion and ion-induced dipole interactions alter their values upon randomization. The contribution that is most dramatically altered — increased — by fixing the molecules is the dipole-dipole part. We quote from Davies: "... predominance of the dispersion energy is a characteristic of non-polar molecules or of the gas phase only. In liquids or solids where the molecules are at much closer distances, the random orientation which reduces the dipole-dipole term to dependence upon r^{-6} in the gaseous state, is far less likely to be maintained. Dipolar molecules can then assume fixed orientations with respect to one another with greatly increased energies of interaction". Thus dipole-dipole, ion-dipole and to a lesser extent, dipole-induced dipole interactions become more important in condensed phases. (For a comparison between gaseous and solid HCl, see[10], p. 166)

The above description disregarded the possible role of quadrupole moments. These are often considered small and negligible. While this is almost certainly so for quadrupole-quadrupole interactions, dipole-quadrupole and ion-quadrupole interactions might be appreciable. Terms like $-C_2/r^8$ and $-C_3/r^{10}$ can be added to equation [12] to take these interactions into account.

The next step is to solve the Schrödinger equation with a Hamiltonian containing the potentials corresponding to all possible types of interaction. This would yield the interaction energies and the charge distribution. The results might exhibit arrangements called donor-acceptor (charge transfer) complexes and hydrogen bonds. These involve transfer of electronic charge between molecules. In what follows we shall often encounter these, especially hydrogen bonds. Their energies are made up by the interactions listed above to which wave mechanics adds its own, more mysterious contribution. At this point, we make only one comment, however: hydrogen bonds are wonderful devices to force dipoles into fixed orientations.

These considerations might appear to be elementary but they have an immediate bearing on our ideas concerning the mechanism of anesthesia.

3 Some Observations on Anesthetic Molecules

It is sometimes said in informal conversations that "almost everything is anesthetic". This statement is not outrageous. There are hardly any molecules that have no anesthetic potency under appropriate conditions, pressure etc. According to a tabulation by Miller and Smith ([2], p. 139), only helium and neon show no anesthetic potency for pressures less than 140 atm. The other rare gases as well as H_2, N_2,

SF_6 or CF_4 do. (Yet, few molecules are more inert than CF_4 or SF_6). As we have already stated, all available evidence indicates that the process of anesthesia involves changes in molecular associations only, not chemical reactions. Then, since "almost everything is anesthetic" and since what "almost everything" can do is interacting with dispersion forces, it may seem to be logical to jump to the conclusion that the forces that are involved with anesthesia are essentially dispersion forces.

General anesthetics are soluble in lipids. Only a few are soluble in water. Furthermore, there is a well known correlation between anesthetic potency and lipid solubility. It is the Meyer-Overton rule that has been known for 80 years to researchers in anesthesia.[12,13]. This relationship was thoroughly studied and re-examined in recent years (See[1]). In its most modern form the lipid solubility or oil/water partition coefficient is plotted against the so-called "righting reflex" taken for a measure of anesthetic potency. It is log 1/p where p is the effective anesthetic pressure in atmospheres required to suppress the righting reflex of mice in half of the experimental animals[2,14,15]. On this relationship are based the "unitary hypothesis" and the "hydrophobic site theory" which state that all general anesthetics act by the same mechanism at the same molecular or sub-cellular sites of the membrane and that the sites are hydrophobic.

The hydrophobic site is usually a non polar group like a hydrocarbon chain. If a molecule contains both non-polar and polar groups then in an aqueous medium it will take up a configuration or associate in such a way that the hydrophobic (non-polar) parts are in contact with one other so as to maximize the number of attractive interactions (mainly by dispersion forces) and that the hydrophilic (polar) parts are turned toward the aqueous phase to participate in hydrogen bonds or other polar interactions. Both enthalpy and entropy considerations are important in this respect and we refer to Nemethy and Scheraga[16] for a detailed discussion of the problem.

That the lipid solubility versus anesthetic potency relationship is not above criticism has been intimated for a number of years by a number of authors. Summaries of the relevant facts and comments are found in the reviews of Halsey[3] and Kaufman[5]. It is only since 1974, however, that the possible importance of polar interactions has become a target of intense discussions. General anesthetics have widely different chemical structures and it has never been possible to classify them on chemical grounds. Xenon, nitrous oxide, ethylene, cyclopropane, ether, chloroform, C_2F_6, SF_6, $CF_3{-}CHCl_2$, $CF_3{-}CHClBr$ (halothane), $CH_3OCF_2CHCl_2$, (methoxyflurane) can all exert anesthetic action. (This aspect will be discussed in more detail in the next section). Looking at the formulas of these different molecules it is hard to believe that they all associate with the same site and with the same type of forces. A series of observations have been made in recent years that substantiate this scepticism.

It has been observed in our laboratory[17-22] that fluorocarbon type anesthetics contaning higher halogens hinder the formation of hydrogen bonds. The latter were of the 0—H———O, N—H———N, N—H———O=C < and S—H———S types. When $CF_3CHClBr$ (halothane), CF_3CHCl_2, or $CF_2Br{-}CF_2Br$, for example, are added to solutions containing alcohols, phenols, amines, amides, thiols engaged in such hydrogen bonds, the ratio of the intensities of the free (non-hydrogen bonded) and associated (hydrogen bonded) infrared OH, NH or

SH stretching bands is altered so that the free bands become more intense and the associated bands less intense. Subsequently we made similar observations on hydrogen bonds involving water, like for example in water-ether systems[23].

At this point an interesting observation can be made. Fluorine atoms are put into anesthetics in order to make the molecules more inert. Instead of entering chemical reactions such molecules only participate in intermolecular associations in the cell membrane, a condition for being used as anesthetics. The most potent anesthetics, however, usually contain a "last" hydrogen, often referred to as the acidic hydrogen. Chloroform, halothane, methoxyflurane are in this category. Molecules containing this C—H bond are, of course, expected to enter polar interactions.

Evidence for this was presented by a number of authors. Davies, Bagnall, Jones and Bell[24, 25], worked out a model to estimate gas/oil phase distribution coefficients for numerous anesthetics through activity coefficients approximated as a function of the dominant intermolecular interactions in the given phases. They found that for halogenated anesthetics this function "reduces to an empiric balance between the compounds' Van der Waals and hydrogen bond donor properties". Hansch et al.[14] who examined the partition coefficients of a number of gaseous anesthetics in the octanol-water system also came to the conclusion that the relative anesthetic potency depends on a hydrophobic and on a polar factor. Di Paolo, Kier and Hall[15] arrived at similar conclusions through connectivity index calculations. Massuda and Sandorfy[22] studied the associations between anesthetics containing the acidic hydrogen and H-bonds of the N—H———O, O—H———O and N—H———O=C types by infrared spectroscopy and demonstrated the formation of C—H———N or C—H———O type hydrogen bonds and the simultaneous "breaking" of a part of the original hydrogen bonds. Brown and Chaloner[26] observed amide-halothane hydrogen bond formation by NMR and Koehler, Curley and Koehler[27] substantiated the role of polar interactions through measuring the NMR spectra of halothane in a number of solvents.

We conclude that there are strong indications to the effect that both non-polar and polar interactions play a role in the anesthetic process. The nature of these interactions will be further examined in subsequent sections.

4 Classification of General Anesthetics

As stated in the preceding section general anesthetics do not belong to a well defined chemical category; they have widely different structures. Actually, attempts to classify general anesthetics on chemical grounds have never been successful. The reason for this appears clearly: since anesthesia is not a matter of chemical reactions but a matter of intermolecular interactions a pertinent classification must be based on the types of interactions that any given anesthetic molecules can participate in. Such a classification has been proposed[28] and will be given here with slight modifications.

1) Anesthetic molecules whose associations are determined essentially by their polarizabilities. This relates to non-polar anesthetics where all the association energy is due to dispersion forces only or to dispersion forces and ion or dipole-

induced dipole interactions with the induced dipole in the anesthetic. Examples are xenon, SF_6, C_2F_6, ethylene, paraffins. While dispersion forces depend to some extent on the ionization potentials too this is not considered essential from the point of view of the proposed classification.

Non-polar and weakly polarizable molecules would be expected to interact mainly with the hydrophobic parts of the membrane lipids or proteins. Non-polar or weakly polar but highly polarizable anesthetics could interact with both hydrophobic and ionic or polar sites.

2) Anesthetics that associate by donor-acceptor (charge transfer) complex formation, either as electron donors (low ionization potential) or as electron acceptors (high electron affinities). There are no confirmed examples of this but fluorocarbon anesthetics containing higher halogens seem to be eligible.

3) Anesthetics with appreciable dipole moments. In addition to interacting with dispersion and induction forces they can enter electrostatic interactions of the ion-dipole or dipole-dipole type. Such molecules are likely to associate with the polar sites in lipids and proteins. N_2O is an example. Anesthetics with no appreciable dipole moment but high bond moments might also be put into this category.

4) A very important special case of 3) are anesthetics that can form hydrogen bonds either as proton donors or proton acceptors. Examples for the former are chloroform or halothane and for the latter ethers.

Some anesthetics like methoxylflurane could function both as proton donors and as proton acceptors.

These different cases will be examined more closely in the following sections. The underlying idea is that a unitary theory of anesthesia can be maintained only in as much as the forces that act involve in all cases only intermolecular interactions. But all the different types of intermolecular interactions have to be taken into consideration.

5 Halogenated Anesthetics Perturb Hydrogen Bonds

It has been known for several years that the degree of association for hydrogen bonded complexes in solution is shifted in favor of the free species in Cl, Br or I containing solvents with respect to hydrocarbon or other, more inert solvents[29-33]. An interesting review has been given by Gomel[32]. It follows that in these halogenated solvents a competing association must exist which is in equilibrium with association by hydrogen bond formation. This competing interaction was thought be of charge-transfer nature[19,31,32] but it now seems to be mainly of the Van der Waals type. Results obtained in our laboratory through infrared spectroscopic measurements extended these observations to fluorocarbons containing higher halogens and led to the suggestion that this hydrogen bond "breaking" property of halogenated molecules is related to their anesthetic potency[19,20].

Our first observations were made on dimethylamine which in solution dimerizes through an N—H———N type hydrogen bond. (Bernard-Houplain and Sandorfy[17]). The self-association of dimethylamine was studied in solution at temperatures ranging from room temperature to liquid nitrogen temperature in two solvents: a) a 1:1 mixture of $CFCl_3$ and methylcyclohexane and b) a 1:1 mixture of $CFCl_3$ and $CF_2Br—CF_2Br$. At room temperature in solvent a) only two species were clearly present in the spectrum of a 0.3M solution: the free species (NH stretching band at 3360 cm^{-1}) and an associated species (3307 cm^{-1}), very probably the dimer. At —190 °C the free band has almost entirely disappeared, the associated band underwent a moderate shift to 3288 cm^{-1}. At higher initial concentrations a third band appeared near 3235 cm^{-1} due to a more highly associated species.

This is normal behaviour. As temperature is lowered the thermal motion is gradually slowed down, the average distance between molecules decreases and conditions become more favorable for association. However, when we measured the spectrum in solvent b) we found a quite different situation. Starting with a solution which at room temperature had the free and associated band at about the same frequencies and with the same relative intensities as before we found, to our surprise, that it is not the associated band whose intensity increases upon cooling but the intensity of the free band. When we reached —190 °C the associated band has almost complety disappeared and the free band become preponderant. Figure 1, compares the spectra of dimethylamine in solvent b) at different temperatures. As to the band of the higher associated species it did not appear at all in solvent b).

This spectacular change in the spectrum is certainly connected with the presence of $CF_2Br—CF_2Br$. N—H———N hydrogen bonds are broken and are replaced by some other type of competing association which is favored under the given conditions. It was first thought that the competing interaction could be charge transfer involving the bromines but subsequent work indicated that this is not very probable. Interactions through dispersion and induction forces seem to be more likely, due to the relatively high polarisability of such bromine containing molecules (see p. 113).

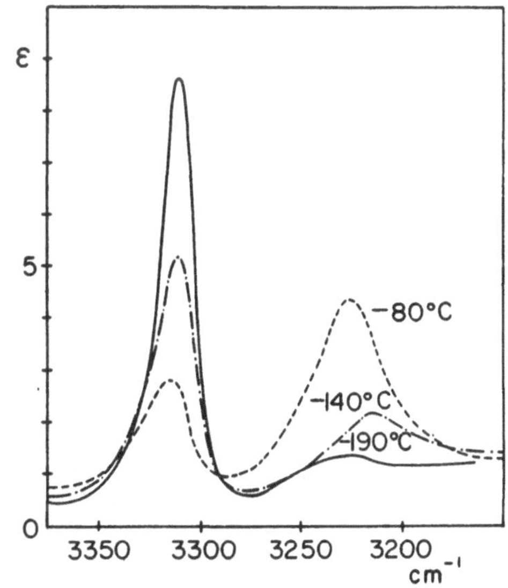

Fig. 1. The NH stretching region of the infrared spectrum of a 1.248M solution of diethylamine in a 1:1 mixture of $CFCl_3$ and $CF_2Br—CF_2Br$ at different temperatures. From M. C. Bernard-Houplain, C. Bourdéron, J. J. Péron and C. Sandorfy, Chem. Phys. Letters *11*, 149 (1971). Reproduced with permission from the North-Holland Publishing Company

The observed effect was, of course, magnified by the use of low temperatures. To a lesser extent it is present, however, at room temperature as well. The possible biological importance of the observed effect has been recognized at an early stage[18]. CF_2Br-CF_2Br (commercially known as Freon 114-B-2) has a weak anesthetic potency[34] but it is not used in clinical practice. However, many similar compounds are and it occurred to us that the hydrogen bond perturbing ability of fluorocarbon anesthetics might be connected with their anesthetic potency. (It is to be noted that, unlike the most potent fluorocarbon anesthetics, CF_2Br-CF_2Br does not contain the "acidic" hydrogen).

Subsequently an extensive infrared spectroscopic study was undertaken in order to ascertain if this hydrogen bond perturbing ability of fluorocarbon anesthetics is of general occurence and if it can be at least qualitatively related to their anesthetic potency. (Di Paolo and Sandorfy,[19,20]). Most of this work was done at low temperatures. It was found that the phenomenon is quite general.

A great number of hydrogen bonded systems were examined with a great number of Cl, Br and I containing fluorocarbons. The hydrogen bonded systems included self-associated aliphatic amines, alcohols, phenols, pyrrole and indole, n-propylthiol[21] and amides. The fluorocarbons were $CFCl_3$, CF_3-CHCl_2, n-C_3F_7Br, CF_2Br_2, CF_2Br-CF_2Br, $CF_3CHClBr$ (halothane), $CF_2Br-CHFCl$, n-C_3F_7I, i-C_3F_7I, CF_2I-CF_2Br. A few examples are given, taken from the original publications of Di Paolo and Sandorfy[19,20].

In Fig. 2 a dilute solution of N-methylpivalamide is shown in two solvents, a 1:1 mixture of $CFCl_3$ and methylcyclohexane and in a 1:1 mixture of $CFCl_3$ and

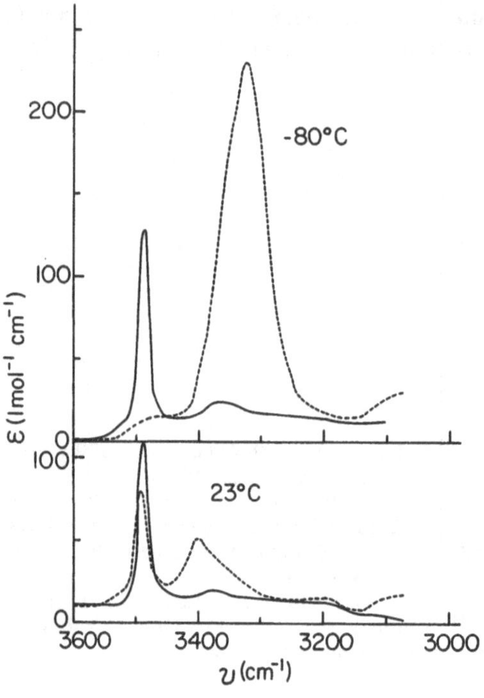

Fig. 2. The NH stretching region of the infrared spectrum of a 0.048M solution of N-methyl-pivalamide in a 1:1 mixture of $CFCl_3$ and methylcyclohexane, ----, and in a 1:1 mixture of $CFCl_3$ and halothane, ——. T. Di Paolo and C. Sandorfy, J. Med. Chem. *17*, 809 (1974). Reproduced with permission from the American Chemical Society

halothane, both at 23 °C and at −80 °C. The *associated* N—H———O=C < band is much weaker, even at room temperature, in the presence of halothane and at −80 °C the difference is spectacular. Fig. 3 shows the OH stretching region of the infrared spectrum of 2,6-diisopropylphenol at −140 °C and −186 °C in a 1:1 mixture of $CFCl_3$ and methylcyclohexane and in the same solvent but with an excess of CF_3—$CHCl_2$ added. Similar observations are made with n-C_3F_7I as the hydrogen bond breaker. (No H in the molecule). In all three examples the hydrogen bond breaking power of the halofluorocarbon is seen to be considerable at low temperatures.

In this study the following general trends have been observed:

1) Perfluoro molecules did not exhibit any observable hydrogen bond perturbing potency.

2) In molecules containing no hydrogen, the order of hydrogen bond perturbing ("breaking") potency was found to be Cl < Br < I.

3) The presence of a hydrogen atom increases greatly the hydrogen bond perturbing potency. As stated previously it also increases anesthetic potency.

4) There exists a parallelism between hydrogen bond perturbing and anesthetic potency for fluorocarbons containing higher halogens.

The effect of the acidic hydrogen having been recognized it was logical to inquire about the mode of its action. Two possibilities come immediately in one's mind. First, the acidic hydrogen might form hydrogen bonds of the C—H———O

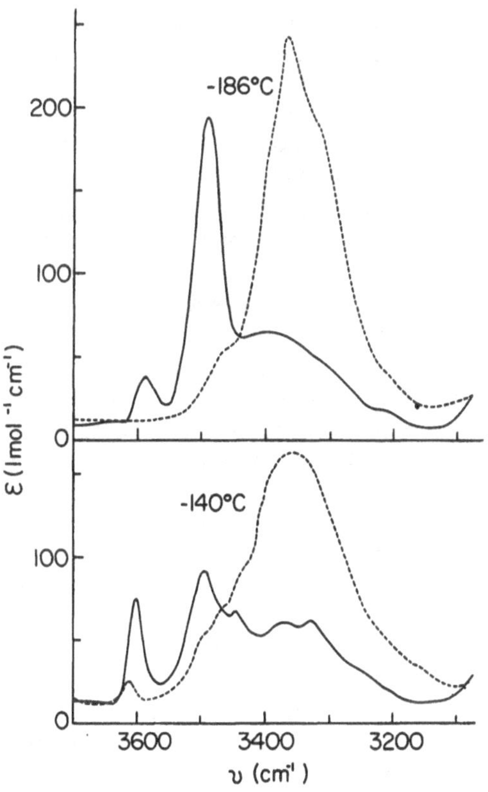

Fig. 3. The OH stretching region of the infrared spectrum of 2,6 diisopropylphenol. 0.054M in a 1:1 mixture of $CFCl_3$ and methylcyclohexane, – – – –, and in the same solution with 1.1M CF_3CHCl_2 added: ———. From T. Di Paolo, C. Sandorfy, J. Med. Chem. *17*, 809 (1974). Reproduced with permission from the American Chemical Society

or C—H———N type as proton donor. Proposals to this effect came about from several quarters and have been mentioned previously[14,15,22,24–27].

Second, the competing interaction might be carried by the higher halogens contained in these molecules. We shall have to examine both possibilities, but in this section only the role of the acidic hydrogen is considered.

Massuda and Sandorfy[22] carried out infrared measurements at temperatures ranging from room to liquid nitrogen temperature on solutions containing self associated secondary amines, amides or alcohols which contain hydrogen bonds of the N—H———N, N—H———O=C < and O—H———O types. An anesthetic containing the acidic hydrogen had been added to the solution: chloroform, halothane ($CF_3CHClBr$), enflurane (CF_2HOCF_2CHFCl), methoxyflurane ($CH_3OCF_2CHCl_2$), among others. The results of these studies can be illustrated on the example of diethylamine. In a 1:1 mixture of $CFCl_3$ and methylcyclohexane considered, in this context, as an inert solvent, the associated N—H stretching band is at 3268 cm^{-1} at 22 °C. On lowering the temperature it gradually increases in intensity and shifts to lower frequencies. At −180 °C it is at 3224 cm^{-1}. A weak free NH stretching band can still be seen at about 3310 cm^{-1}. With a similar solution but with 1.4M chloroform added, the growth of the associated band on lowering the temperature is much slower. The free band is quite intense at all temperatures. This clearly demonstrates the hydrogen bond "breaking" property of chloroform. It is then natural to look at the behavior of the C—H stretching band of chloroform. Actually deuterated chloroform was used in order to avoid interference from other CH bands but the same results were obtained with $CHCl_3$ and $(CD_3)_2NH$ in a 1:1 mixture of $CFCl_3$ and methylcyclohexane —d_{14}. The free $CDCl_3$ stretching band has its maximum at 2250 cm^{-1}. The associated band was found to be at 2187 cm^{-1} at room temperature. Its intensity increases gradually on lowering the temperature. It broadens at its low frequency side due probably to the appearance of more highly associated species. This clearly demonstrates the fact that as N—H———N type hydrogen bonds are broken, C—H———N type hydrogen bonds are formed and an equilibrium is reached corresponding to the given temperature and concentrations.

Similar results have been obtained with self-associated N,N-dimethylacetamide and tertiary butanol with the above mentioned anesthetics. This lends weight to the suggestion that the interaction competing with the existing hydrogen bonds in these systems is the formation of a different hydrogen bond which is of the C—H———N or C—H———O type. Evidence for this has been found by Brown and Chaloner[26] through nuclear magnetic resonance studies in halothane + N-methylpyrrolidone, (1—H)-undecafluorobicycloheptane + N-methylpyrrolidone and similar systems. Martire et al.[35] used both NMR and gas chromatographic methods to examine the association between chloroform and an ether, a thioether and an amine. According to their results chloroform interacts with the latter molecules both through hydrogen bond formation as a proton donor and through the chlorine atoms.

In order to ascertain if hydrogen bonds involving water can be perturbed by anesthetics we measured the infrared spectra of the water +2-methyltetrahydrofuran system at various temperatures and then added chloroform or halothane to it.[23] 2-methyltetrahydrofuran can be cooled down to liquid nitrogen temperature where it sets to a glass. It dissolves water to some extent.

In our experiments the concentration of water was about 0.04M. In the absence of perturbers only weakly associated OH stretching bands are seen in the spectrum (Fig. 4), at about 3575 and 3490 cm^{-1} with some shoulders. At low temperatures more highly associated species become preponderant; at —190 °C the broad, complex band has its center of gravity between 3300 and 3200 cm^{-1}. In presence of chloroform or halothane (Fig. 5) the center of gravity moves to about 3500–3400 cm^{-1} showing that a part of the hydrogen bonds was broken and the less highly associated species are favored.

While the generality of the hydrogen bond perturbing property of anesthetics could be amply demonstrated the above described infrared studies were far from physiological or clinical conditions. In order to magnify the effects low temperature measurements had to be applied in most (but not all) of our work. This was due to the relative weakness of the free OH and NH bands with respect to the associated bands. This, however, applies only to the fundamentals of the respective vibrations. The free/associated ratio is different for the overtones of the same vibrations[36, 37] and it also varies from one associated species to another[38]. Overtones of the stretching vibrations of hydrogen bonded OH and NH groups are usually very weak contrasting the well known strength and breadth of their fundamentals. An interpretation of this intriguing phenomenon has been proposed[39]. It is connected with the possibility that mechanical and electrical anharmonicities make contributions of opposite sign to the intensity and, in the case of strongly polar vibrators, can largely cancel each other. It happens that the free band of the fundamental is too weak and the free overtone is too strong for a study of the effect of anesthetics on a given hydrogen bond equilibrium to be performed. It has been found, however, (Trudeau et al.[40]) that for certain combination bands, the intensities of the free and associated bands are just in the right proportion. These are

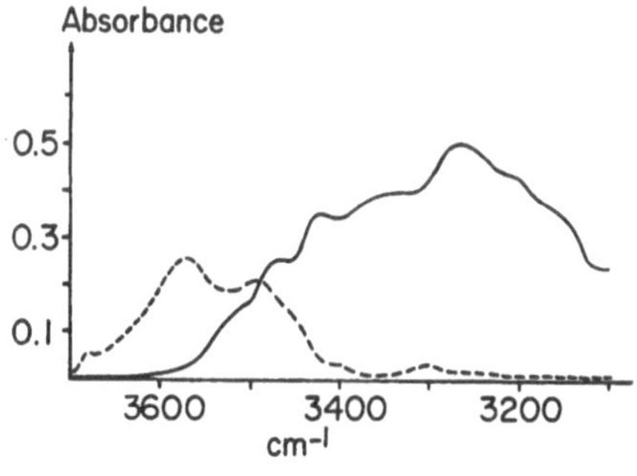

Fig. 4

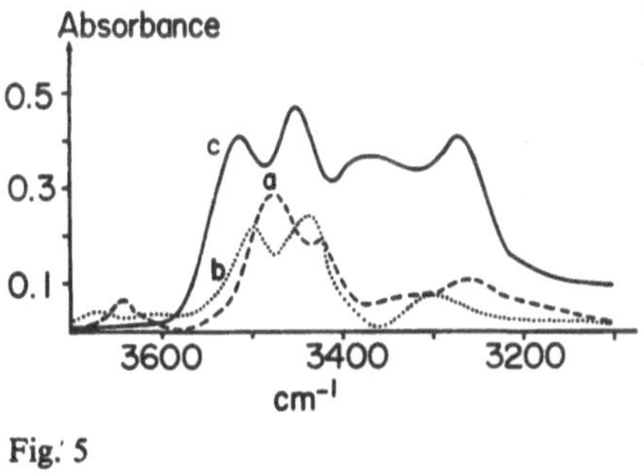

Fig. 5

Fig. 4. The OH stretching region in the infrared spectrum of a 0.04M solution of water in 2-methyltetrahydrofuran. - - - - : 22 °C; ——— —190 °C. From A. Nagyrevi and C. Sandorfy, Can. J. Chem. *55*, 1593 (1977). Reproduced with permission from the National Research Council of Canada

Fig. 5. The OH stretching region in the infrared spectrum of a) 0.03M water + 1.02M chloroform; b) 0.04M water + 0.87M halothane; c) 0.04M water + 0.60M tetramethyl-urea. All in 2-methyltetrahydrofuran at —190 °C. From A. Nagyrévi and C. Sandorfy, Can. J. Chem. *55*, 1593 (1977). Reproduced with permission from the National Research Council of Canada

the combinations involving one quantum of the OH or NH stretching and one of the in-plane bending vibrations. The bands are located between 5300 and 5200 cm^{-1} for the OH and 5000 and 4900 cm^{-1} for the NH for the systems that were examined. These bands made it possible to study the effect of anesthetics on hydrogen bonds at room temperature.

Three model systems were chosen: water and dioxane, self associated N-methyl and N-ethylacetamide and self associated tertiary butanol. The hydrogen bonds are of the O—H–––O (ether), N—H–––O=C and O—H–––O (alcohol) types, respectively. To solutions containing these one of the following perturbers was added: $CFCl_3$, CF_2Br—CF_2Br, CH_2Cl_2, CF_3CHCl_2, $CHCl_3$, CF_3—CHClBr, (halothane) and $CH_3OCF_2CHCl_2$ (methoxyflurane). Care has been taken to have the same molar concentrations for all solutes in a given series. The qualitative result is illustrated in Fig. 6 on the example of self associated N-ethylacetamide. The free NH band has its maximum at 4978 ± 5 cm^{-1} and the associated band at 4905 ± 10 cm^{-1}. The solutions whose spectra are shown in Fig. 6 contained equimolar amounts of methylcyclohexane (E), $CFCl_3$ (F), CH_2Cl_2 (G) and $CHCl_3$ (H). The anesthetic potency of E and F is very weak, that of G and H is much stronger. Accordingly the free band is much more intense in the spectra of G and H showing that many of the N—H–––O=C hydrogen bonds were broken. Table 1 contains some numerical results for this system. The intensity of the free band increases and that of the

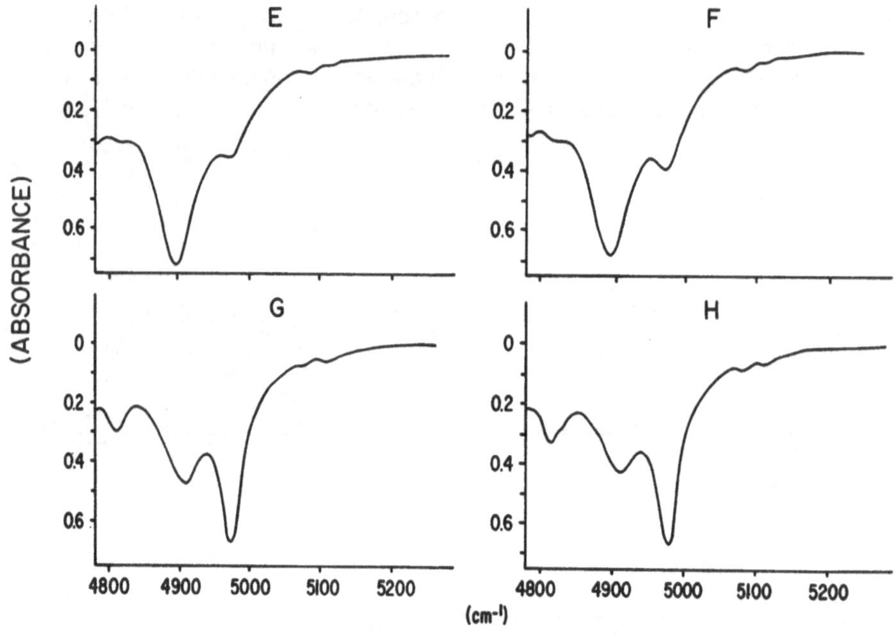

Fig. 6. Changes in the relative intensities of the free and associated NH stretching + NH in plane bending combination band of self-associated N-ethylacetamide. The solutions contained equimolar amounts of one of the following "anesthetics": E, C_7H_{14}; F, $CFCl_3$, G, CH_2Cl_2; H, $CHCl_3$. From G. Trudeau, K. C. Cole, R. Massuda and C. Sandorfy, Can. J. Chem. *56*, 1681 (1978). Reproduced with permission from the National Research Council of Canada

Table 1. Computed relative band areas for the NH stretching + in-plane bending combination. (The band areas are on different relative scales for each system and should only be compared within a given column)[40, 66]

Anesthetic	log 1/p	Band Area self-associated N-ethylacetamide free NH	Band Area self-associated N-ethylacetamide ass. NH
C_7H_{14}	~0	37.78	46.68
$CFCl_3$	0.82	37.89	45.42
CH_2Cl_2	1.52	39.19	42.15
CF_3CHCl_2	1.57	40.90	—
$CHCl_3$	2.08	41.73	33.60
$CF_3CHClBr$	2.11	42.95	30.75
$CH_3OCF_2CHCl_2$	2.66	44.70	29.20

associated band decreases with increasing anesthetic potency. The parallelism is evident. Similar results have been obtained for the other two model systems, all at room temperature[40].

In all three systems the most potent anesthetics, chloroform, halothane and methoxyflurane were also the most potent hydrogen bond breakers. Halogenated anesthetics that contain acidic hydrogen are more potent than those which contain none. As shown previously some of the latter, like $CF_2Br—CF_2Br$, become powerful hydrogen bond breakers at low temperatures but at room temperature they have only a weak effect. $CFCl_3$ is even weaker and can be considered inert for many purposes.

The question still remains if the acidic hydrogen is the only important associative factor in the molecules that contain one or if the halogens too play an important role. Furthermore we have to examine the nature of possible intermolecular interactions in the case of molecules containing no acidic hydrogen[41].

From this point of view the case of CH_2Cl_2 seems to be of special interest. In this molecule there are two hydrogens whose tendency to form hydrogen bonds is somewhat doubtful. Then the question as to whether under given circumstances it enters associations with the hydrogens or with the chlorines appears to be legitimate. In the studies described above, CH_2Cl_2 exhibited a normal behavior with two of our model systems, self-associated amides and alcohols: its hydrogen bond breaking ability was in line with its anesthetic potency. With the water-dioxane system, however, it was found to break more hydrogen bonds than would be expected. This molecule was subjected to further investigation.

5.1 The Case of CH_2Cl_2

The vibrational spectrum of CH_2Cl_2 has been the subject of a number of publications[42–45]. In the vapor phase the asymmetrical CH stretching frequency is absent while the symmetrical stretching frequency gives a well defined doublet at 2995 cm^{-1}. Evans and Lo[43] measured the intensities of these bands in mixed solvents containing proton acceptors at a variety of concentrations. They noted the rather different behavior of the two bands. For example, in dilute solutions in acetone —d_6/CCl_4 the asymmetrical mode (3050 cm^{-1}) increases rapidly in intensity with increasing acetone concentration while the intensity of the symmetrical mode (2980 cm^{-1}) remains practically unchanged. Curiously, the *frequencies* of the two C—H stretching bands are hardly affected by changes in the composition of the solvent. However, Oi and Coetzee[44] observed that the two C—Cl stretching vibrations underwent changes in frequency parallelling the proton acceptor ability of the solvent. The asymmetrical C—Cl frequency which is at 759 cm^{-1} for the vapor shifted to 734 cm^{-1} in 1,4—dioxane, for example, while the symmetrical C—Cl frequency moved from 725 to 702 cm^{-1}. So this is a case in which hydrogen bond formation affects the *intensities* of the motions directly implying the proton donor group but not their frequencies while there are indirect effects observable on other frequencies. Whether we have to call this interaction a hydrogen bond formation or not is probably a matter of semantics but the interactions exists. Evans and Lo[43] have shown that bond-dipole moment changes in the C—Cl bonds due to the C—H motions $\frac{\partial\mu(C{-}Cl)}{\partial r(C{-}H)}$ linked to changes in hybridization on the common carbon atom give rise to appreciable cross terms which lead to a near annihilation of the intensity of the asymmetrical C—H stretching vibration.

Thus we are led to admit that in certain cases hydrogen bond formation can be monitored by intensity changes only but the question still remains if in actual associations the Cl atoms or C—Cl bonds are directly perturbed or only suffer the indirect effect of the perturbation of the H atoms or C—H bonds. In order to clarify this we first carried out low temperature and then room temperature work on CH_2Cl_2 and similar molecules.

First the spectrum of CD_2Cl_2 was recorded in the mixed solvent diethyl ether and methylcyclohexane (2:3) at 30 °C intervals from room temperature to liquid nitrogen temperature. Deuterated dichloromethane was used in order to avoid interference from the C—H vibrations of the solvent molecules. The asymmetrical and symmetrical C—H stretching vibrations have their maxima at 2298 and 2194 cm^{-1} respectively at room temperature and at 2297 and 2192 cm^{-1} at —180 °C. So practically no frequency change is observed throughout the about 200 °C temperature interval. Intensity changes are observed, however. They are quite pronounced for the asymmetrical band.

Next, we measured the C—Cl bands for CH_2Cl_2 in methylcyclohexane throughout the same temperature range and observed no frequency changes. However, in the mixed solvent diethyl ether/methylcyclohexane the asymmetrical C—Cl vibration shifted from 740 to 727 cm^{-1} and the symmetrical one from 706 to 696 cm^{-1}. These are slight but significant differences. Also the asymmetrical band underwent a large increase in intensity.

In stronger electron donors like pyridine in a mixture of $CFCl_3$ and methylcyclohexane—d_{14} slight shifts have been observed: 4 cm^{-1} for the asymmetrical C—H stretching vibration (3051 to 3047 cm^{-1}) and 10 cm^{-1} for the symmetrical one (2985 to 2975 cm^{-1}) in going from room to liquid nitrogen temperature. At room temperature only very slight frequency changes were observed but with significant intensity changes. Similar results were obtained in dimethyl-sulfoxide. This shows that strong proton acceptors are able to force frequency changes on the C—H vibrations of CH_2Cl_2 which then appears as a weak proton donor.

For the sake of comparison we measured the C—Cl bands of CH_3CHCl_2 and $(CH_3)_2CCl_2$ in the 2:3 mixture of diethylether and methylcyclohexane. According to Gasanov[46] the asymmetrical and symmetrical C—H stretching bands of $(CH_3)_2CCl_2$ are at 657 and 553 cm^{-1} respectively, at frequencies significantly lower than in CH_2Cl_2. For a 0.23M solution the upper frequency shifted to 651 cm^{-1} from room temperature to —170 °C while the band of lower frequency did not change. The asymmetrical C—Cl frequency of CH_3CHCl_2 shifted from 690 to 682 cm^{-1}, slighly more. This seems to show that the larger shifts of the C—Cl bands observed for CH_2Cl_2 are due to the indirect affect of associations affecting primarily the hydrogens, at least in polar media. Direct solvent effects on the Cl are weak but as the comparison with $(CH_3)_2CCl_2$ and CH_3CHCl_2 shows they do exist. Then it seems probable that in the absence of polar sites CH_2Cl_2 itself could associate with its chlorines. Since it is a small molecule several such interactions could affect a given hydrophobic chain and the interaction energy might become significant. While the extents of direct and indirect effects on the C—Cl frequency are difficult to estimate the shifts observed for $(CH_3)_2CCl_2$ show that direct associations involving the chlorines can occur.

6 Molecules Containing the Acidic Hydrogen: The Hydrogen and the Halogens

While the brunt of intermolecular interactions is expected to be carried by the acidic hydrogen for these molecules, it is still of importance to gain knowledge on the possible role of the halogens. That the acidic hydrogens enter hydrogen bonds is, of course, well known[30,45].

First, we measured the IR spectrum of a dilute solution of $CDCl_3$ in the 2:3 mixture of diethylether and methylcyclohexane. At room temperature the C—D stretching band is at 2243 cm^{-1}. At —180 °C it is at 2233 cm^{-1}, a shift of 10 cm^{-1}. At the same time the degenerate C—Cl stretching vibration shifted from 757 to 749 cm^{-1} but the symmetrical one (665 cm^{-1}) remained unchanged. (For the assignments see Herzberg[47]). Oi and Coetzee[44] argue that the variation of the C—Cl frequency is an indirect result of hydrogen bond formation by the molecule's acidic hydrogen. The CH bands of CF_3CHCl_2 (commercially known as freon-123) and $CF_3CHClBr$ (halothane) behave in a way similar to chloroform. The C—H stretching band of CF_3CHCl_2 is at 2975 cm^{-1} (for assignments see Nielsen et al.[48]) at room temperature and at 2953 cm^{-1} at —185 °C. For $CF_3CHClBr$ the respective frequencies are 2976 and 2957 cm^{-1}. These molecules are more acidic than chloroform and this is in keeping with the larger shifts that were observed. The C—Cl bands, 827 and 763 cm^{-1} for CF_3CHCl_2 and 808 cm^{-1} for halothane did not shift on cooling to low temperature, nor did the C—Br band of halothane. This might appear surprising since the Cl and Br are attached to the same carbon as the acidic hydrogen just as in $CHCl_3$ or CH_2Cl_2. It could have been expected that the hydrogen bond formation has an indirect effect on the C—Cl and C—Br bonds. It seems then that the presence of the CF_3 group makes the C—H bond more acidic and the electrons in the C—Cl and C—Br bonds more difficult to perturb. In strong proton acceptors like pyridine and dimethylsulfoxide a slight decrease in the C—Cl frequencies was observed. The global result of these investigations is that the effect on the C—Cl force constant is very likely indirect.

In both our room and low temperature studies we were led to the conclusion that while in molecules containing the acidic hydrogen the interaction is mainly through that hydrogen and the effect on the C—Cl is indirect, the chlorines can enter into direct interactions and in the absence of the acidic hydrogen these are the only ones that remain.

For halothane (CF_3—CHClBr), the C—Cl and C—Br vibrations are hardly affected by pyridine at room temperature. In accordance with the results of our low temperature study the C—H bond becomes more acidic and the C—Cl and C—Br bonds less associable because of the presence of the fluoromethyl group.

Methoxyflurane is a very powerful anesthetic and constitutes an interesting problem. It has three associative sites:

$CH_3OCF_2CHCl_2$

the acidic hydrogen, the chlorines and the ether group. But for the latter the molecule resembles CF_3—$CHCl_2$. Actually the results we obtained are similar. In a 0.3M solution in 2:3 ether/methylcyclohexane mixture the C—H stretching band is at 2990 cm^{-1} at room temperature and it shifts by 12 cm^{-1} on cooling down to liquid nitrogen temperature. This is less than the shift for halothane or CF_3—$CHCl_2$. But it indicates hydrogen bond formation of the C—H———ether type.

Through comparison with other similar molecules we can locate the two C—Cl stretching vibrations at 811 and 765 cm^{-1}. On cooling to liquid nitrogen temperature the upper band shifted by 5 cm^{-1} and the lower band did not shift. As before we see that the presence of the fluorines makes the chlorines less associable.

As to the oxygen we know from photoelectron studies on fluorinated ethers[49)] that the ionization potential of the lone pair which is near 9.6 ev for diethylether, for example, moves to much higher energies (about 13 ev) when the oxygen has a fluorinated carbon as a neighbor. Thus the availability of the oxygen lone pair for associations in which it should donate electronic charge must be very low. It is not expected to form hydrogen bonds as a proton acceptor. It might, perhaps, link up with electron donors like negatively charged groups in the membrane as an electron acceptor. This is, of course, speculation. However, in view of the relatively modest shift of the C—H vibration and the very slight one of the C—Cl we need an additional site of association in order to understand the strong anesthetic power of methoxyflurane.

$CFCl_3$ has only a very weak anesthetic potency. In keeping with this we found no frequency shift of the C—Cl bands on cooling from room to low temperatures. (One of the bands, the degenerate one, seems to be affected by Fermi resonance.)

6.1 The Case of the —CHF_2 Group

An interesting case is that of molecules which would be perfluoroparaffins except for one "last" hydrogen like 1-H-perfluoroheptane. The question arises if this hydrogen is an "acidic" one? Is it able to form hydrogen bonds like the hydrogen of chloroform?

Allerhand and Schleyer[30] measured the IR spectrum of this molecule in CCl_4 and found two C—H bands, at 3004 and 2978 cm^{-1}. We have measured it in a 2:3 mixture of diethylether/methylcyclohexane and found that the C—H band shifted to *higher* frequencies on cooling. The maximum of the band was at 3005 cm^{-1} at 22 °C and at 3015 cm^{-1} at —100 °C. The band has a pronounced shoulder at its low frequency side at about 25 cm^{-1} from the maximum. They are not a pair of free and associated bands, however. Their relative intensities remained unchanged in wide ranges of temperature and concentration and even in the gas phase. It is believed that they correspond to two conformers. We found a similar shift to *higher* frequencies in presence of acetone —d_6 at room temperature. Association, whether provoked by cooling to lower temperatures or by increasing the concentration of the proton acceptor shifts the C—H band of CHF_2 to higher frequencies. An intuitive explanation of this could consist in assuming that in CHF_2 repulsive interactions predominate above the attractive ones in the formal hydrogen bond F_2C—H——:O. This would mean that the fluorines when in second position as in CF_3—$CHCl_2$ make the "acidic hydrogen" more acidic but when they are at the same carbon like in —CHF_2 they make it less acidic. This is not altogether clear and would warrant a detailed theoretical study.

A frequently used anesthetic containing such groups is enflurane, ClFHC—CF_2—O—CHF_2. The two hydrogens have different environments; the one having both Cl and F at its carbon is expected to be more mobile. We recorded the spectrum of 0.4M enflurane in CCl_4 and then in 1:5 and 1:10 mixtures of acetone —CCl_4, pyridine—CCl_4 and dioxan—CCl_4. The dilution was made in such a way that the concentration of enflurane was always the same, only the concentration of the electron donor varied. In CCl_4 enflurane has two C—H stretching bands, at 3027 and 2995 cm^{-1} respectively, the one at higher frequency being more intense. In 1:10 acetone—CCl_4 the bands are at 3023 and 3003 cm^{-1} and the band at lower frequency becomes the more intense one. If the concentration of acetone is increased, the 3003 cm^{-1} band keeps increasing in intensity and the 3023 cm^{-1} band becomes a shoulder. The two bands found at 3025 and 2997 cm^{-1} in 1:10 dioxan —d_8—CCl_4, behave in the same way. When the electron donor is pyridine—d_5 the 3025 band is unchanged but the other band is at lower frequences, 2973 cm^{-1}, and has a shoulder at its high frequency side. The 2973 band becomes gradually more intense with increasing pyridine concentration and shifts slowly to lower frequencies. We have also recorded the spectrum of enflurane in a ternary mixture of acetone —d_6, pyridine —d_5 and CCl_4 (Fig. 7). Three bands are found, at 3023, 3002 and 2975 cm^{-1} respectively. Clearly they are characteristic of the relatively "free" band in CCl_4, to the enflurane-acetone and enflurane-pyridine interactions, in this order. On the basis of these results we assign the 3025 cm^{-1} band which is practically insensitive to the presence of donors, to the —CHF_2 group and the bands of lower frequency, to the ClFCH— group. Allerhand and Schleyer observed previously[30] that the band in 1-H-perfluoroheptane does not shift upon addition of pyridine. Since the latter molecule only contains the —CHF_2 group this is in agreement with our assignments.

Summing up, then, enflurane might enter into repulsive interactions through its

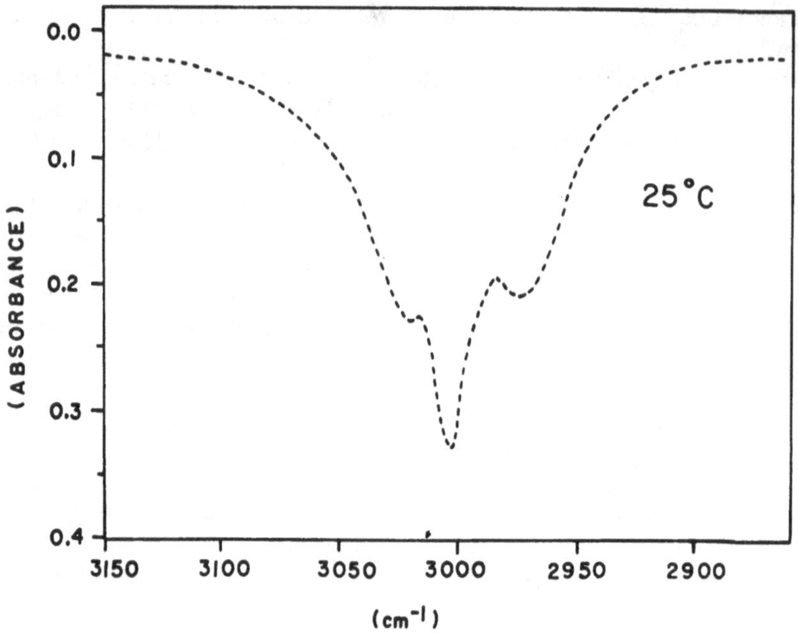

Fig. 7. The C—H stretching region in the infrared spectrum of 0.40M enflurane in a 1:1:10 mixture of acetone-d_6, pyridine-d_5 and carbon tetrachloride at room temperature

Table 2. Chemical shifts of some anesthetic molecules determined with a JEOL JNM 4-H-100 spectrometer: — in absence of a proton acceptor; * with pyridine as proton acceptor

Compound	Solvant	T °C	proton acceptor	Shift (p.p.m)
CH_2Cl_2	CS_2	25	—	4.52
CH_2Cl_2	CS_2	25	*	4.53
CH_2Cl_2	CS_2	−50	*	4.53
CH_2Cl_2	CS_2	−100	*	4.54
CH_2Cl_2	CCl_4	25	—	4.53
CH_2Cl_2	CCl_4	25	*	4.52
$CHCl_3$	CS_2	25	—	4.72
$CHCl_3$	CS_2	25	*	4.74
$CHCl_3$	CS_2	−50	*	4.79
$CHCl_3$	CS_2	−100	*	4.84
$CF_3CHClBr$	CCl_4	25	—	4.57
$CF_3CHClBr$	CCl_4	25	*	4.61
$CF_3CHClBr$	CS_2	25	—	4.57
$CF_3CHClBr$	CS_2	25	*	4.62
$CF_3CHClBr$	CS_2	−100	*	4.75

$-CHF_2$ group and attractive ones with stronger bases through its ClFHC— group. To this one might add the possible electron acceptor role of the oxygen, like in the case of methoxyflurane.

All results described so far have been obtained through infrared spectroscopy. The proton NMR spectra of CH_2Cl_2, $CHCl_3$, CF_3CHCl_2, $CF_3CHClBr$, $CH_3OCF_2CHCl_2$ and CH_3CHCl_2 (0.2M) were also measured. CCl_4 and CS_2 were used as solvents and pyridine as the proton acceptor. In NMR spectra hydrogen bond formation causes generally a shift to lower fields of the proton signal. Since the lifetime of the hydrogen bonded complex is usually too short at the NMR time scale, at normal temperatures we cannot distinguish the free from the associated signal, we can only observe a mean chemical shift. Table 2 shows the results obtained at 22 °C and at −100 °C (in CS_2). As expected associations become stronger at low temperatures causing an additional shift to lower fields.

The shifts we have measured are small but reproducible. CH_2Cl_2 gave no shift even at −100 °C in pyridine $-CS_2$ mixtures, nor did CH_3CHCl_2. The more acidic compounds gave shifts parallelling those of the infrared C—H stretching vibration.

7 Predicting Cases of Antagonism Between General Anesthetics

It can be taken for a fact that anesthetics exert their action in the nerve cell membrane through perturbing molecular associations therein. Since this is so there are two predictable ways in which anesthetics can become antagonists of each other: first, they might compete for the same associative site in the membrane and second, they might associate with each other, in this way precluding association with sites in the membrane. In order to throw some light on these conditions we have undertaken an infrared spectroscopic study which consisted in looking at the stretching-bending NH combination bands of self-associated N-ethylacetamide in solutions containing two different anesthetics. (See Sect. 5.) The relative intensities of the free and associated bands due to this combination were measured. The results obtained with the halothane-ether mixture will be described in some detail. The bands are located in the near infrared; the free band is at about 4970 and the associated band to about 4910 cm^{-1}. First a 0.27M solution of N-ethylacetamide in CCl_4 was prepared to which 0.04 mole of halothane was added. Under these circumstances the free band is by far the more intense. To the contrary if, instead of halothane we add 0.04 mole of methylcyclohexane to the solution the associated band is prominent and overlaps strongly with the free band. This illustrates well the hydrogen bond breaking property of halothane. With 0.04 mole of diethylether the situation is different: the associated band is even more intense than with methylcyclohexane. This can only mean that diethylether associates (as a proton acceptor) with the free N-ethylacetamide molecules. The free band appears only as a shoulder. Next we diminished the concentrations of the anesthetics to half (0.02 mole) of their previous value. With halothane the free band was still the more intense but it lost some of its intensity in favor of the associated band. With ether the associated band become somewhat weaker and the free band (shoulder) more pronounced showing that some of the N—H——O=C < bonds were broken and were only partly replaced by N—H———O < bonds. (Fig. 8).

Now if we add both 0.02 mole of halothane and 0.02 mole of ether to the original solution of the amide in CCl_4 we find that the associated band is much more intense than the free band and that the observed effect is *not* the mean of the effects exerted by the two anesthetics applied separately. It appears that in the presence of ether, halothane breaks a lesser number of hydrogen bonds between the amide molecules than it would by itself at the same concentration. Thus, we can predict that ether and halothane are antagonists.

Ether and halothane are two anesthetics that cannot act through the same mechanism. Halothane was seen to break hydrogen bonds of the N—H———O=C < type by forming hydrogen bonds of the C—H———O=C < type as a proton donor. Ether with its oxygen lone pair can form hydrogen bonds of the N—H———O < type as a proton acceptor. Therefore they are not expected to compete for the same site of association. Now, as has been shown previously ether and halothane associate with each other through C—H———O < hydrogen bonds. Thus while the above results show clearly the antagonism between ether and halothane the manner in which they hinder each other's action is more difficult to pinpoint. Ether could associate with the amide molecules made "free" by the action of halothane or halothane and ether might simply associate between themselves thus preventing each other from associating with the amide. While the distinction cannot be made on spectroscopic ground, the first alternative seems to be very unlikely to apply in the cell membrane. Indeed, if halothane opens a hydrogen bond whereby perturbing the conformation of a constituent of the membrane or an ion channel, the formation of a new hydrogen bond with ether cannot be expected to restore the original order or conformation.

Similar results were obtained with the chloroform-ether-amide system. As chloroform was found to be a somewhat less efficient hydrogen bond breaker than halothane, one would expect a free/associated ratio somewhat less favorable to the free species than in the case of the halothane-ether mixture. This is actually what has been observed. The system is another case of predictable antagonism. The dioxane-halothane-amide system also yielded similar results.

Next we examined the case of the chloroform-halothane-N-ethylacetamide mixture. The effect of chloroform and halothane on the free/associated ratio for the N—H———O=C < hydrogen bonds was found to be additive. Since these two anesthetics act in the same way by their acidic hydrogen this had to be expected.

In order to couple a proton acceptor less efficient than ether to halothane we measured the spectra of the methylsulfide-halothane-N-ethylacetamide system. It was actually found that the thioether hinders the hydrogen bond breaking action of halothane less than the ether; the free/associated ratio is in favor of the free species more than with ether. Ethylsulfide was also tried. In general while thioethers are antagonists of halothane or chloroform they are less so than ethers.

This section might be concluded with the following generalizations which apply at least to relatively low concentrations when not all associative sites of a given type are occupied in the membrane.

1) The effects of anesthetics that associate in the same way should be additive. This applies to all anesthetics whose main associative element is the acidic

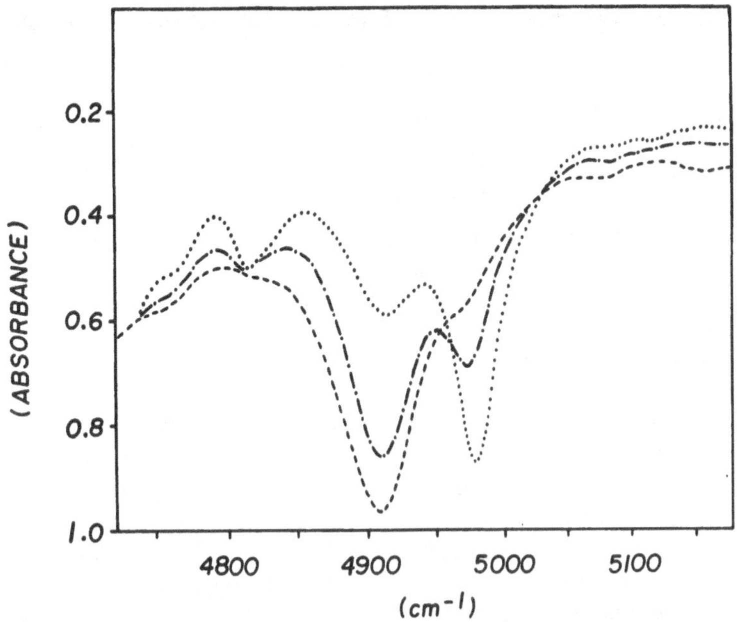

Fig. 8. The NH stretching + NH in plane bending combination band of 0.27M N-ethylacetamide in carbon tetrachloride.

.....: with 0.04 mole of halothane
----: with 0.04 mole of ether
—.—.: with 0.02 mole of halothane and 0.02 mole of ether

hydrogen. The same is expected for anesthetics that associate only through dispersion forces. The combined effect of one anesthetic containing an acidic hydrogen and one that associates with dispersion forces only is likely to be additive as well.

2) Proton acceptors like ethers and thioethers should be antagonists of anesthetics whose main associative element is the acidic hydrogen. This applies to ether-chloroform or ether-halothane mixtures.

8 Polarizabilities of Anesthetic Molecules

When interactions of non-polar molecules are dealt with the most important quantity is molecular polarizability (α). The same applies to inductive Van der Waals interactions. (See Sect. 2). Therefore a search for a relationship between α and anesthetic potency (represented here by log 1/p) is of interest.

The mean polarizabilities ($\bar{\alpha}$) of a number of molecules with anesthetic potency have been determined, others were compiled from the literature. The refractive indices (n) were measured on a precision goniometer mounted with a prism making it possible to select wavelengths and the $\bar{\alpha}$ were computed from the Lorentz-Lorenz equation:

$$\frac{n^2 - 1}{n^2 + 2} = \frac{4}{3}\pi\frac{\varrho}{M}N\bar{\alpha}\,, \tag{13}$$

where ϱ is the density in g/cm^3, M is the molecular weight in g and N is the Avogadro number. The refractive index for most of the $\bar{\alpha}$ given in Tables 3 and 4 was measured at the sodium D line. For halothane, enflurane, methoxyflurane, CF_2Cl—CH_2Cl, $CFCl_2$—CF_2Cl they were measured at a number of wavelengths and extrapolated to infinite wavelength. The $\bar{\alpha}$ values obtained in this way differed by less than 3% from the values computed with n_D. All values were measured at 25 °C unless indicated otherwise.

The α values are characteristic of given molecules in a global way and are not necessarily ideal quantities for the study of interactions in a given geometrical arrangement; comparisons between different molecules also require precaution in order to be meaningful. For constructing Table 3 first a few reference molecules were chosen. These are molecules containing carbon, fluorine and hydrogen only. The α values of other molecules containing the same number of bonds are then compared with that of the reference molecule and between themselves. It is helpful first to have a look to the related bond polarizabilities. These were determined many years ago by Vogel et al.[50)].

Bond	Bond polarizability
C—C	$0.05135 \cdot 10^{-23}$ cm^3
C—H	0.06640
C—F	0.05777
C—Cl	0.25730
C—Br	0.37210
C—O	0.06062

(computed with n_D)

The values in brackets in Table 3 have not been determined experimentally so we computed their approximate values from the bond polarizabilities supposing additivity.

In the first group CF_4 was taken for reference. When the fluorines are successively replaced by chlorines $\bar{\alpha}$ increases gradually. When they are replaced by bromines the increase is larger. It is seen despite the limited number of log 1/p values that are available that the increase of $\bar{\alpha}$ is paralleled by an increase in anesthetic potency when the data are available. For the second group the reference molecule is CHF_3. Replacement of the F by Cl again gradually increases $\bar{\alpha}$ and log 1/p; taking Br instead of Cl causes further increase in both. It is then natural to think that in both groups the favored intermolecular interactions are Van der Waals interactions of the dispersion and induction type. The latter is expected to be important only if the "partner" is an ion or a strongly dipolar molecule.

Table 3. Polarizabilities and anesthetic potencies of halogen derivatives of methane

	$\bar{\alpha} \cdot 10^{23}$ cm^3	log 1/p
CF_4	(0.231)	
CF_3Cl	(0.431)	
CF_2Cl_2	0.656	0.40
$CFCl_3$	0.842	0.82
CF_3Br	0.578	(1.1)*
CF_2Br_2	(0.860)	
$CFBr_3$	1.190	
CHF_3	(0.240)	
CHF_2Cl	0.460	0.80
$CHFCl_2$	0.650	(1.2–1.5)*
$CHCl_3$	0.845	2.08
CHF_2Br	(0.554)	
$CHFBr_2$	0.874	
$CHBr_3$	1.182	
CHFClBr	0.758	(1.6–1.9)*
$CHCl_2Br$	0.959	
$CHClBr_2$	1.070	
CH_2F_2	(0.248)	
CH_2FCl	(0.448)	
CH_2FBr	(0.563)	
CH_2Cl_2	0.651	1.52
CH_2Br_2	0.868	
CH_2ClBr	0.758	
Xe	0.4	
Cyclopropane	0.57	
Ethylene	0.43	
N_2O	0.31	

* estimated

More interesting is the comparison between molecules having the same number of Cl or Br atoms. While $CFCl_3$ and $CHCl_3$ have practically the same polarizability their anesthetic potencies differ dramatically. The same applies to the CF_2Cl_2—$CHFCl_2$ and CF_3Cl—CHF_2Cl pairs. The difference goes decreasingly when the number of fluorines increases. These comparisons illustrate the importance of the acidic hydrogen and polar interactions for the stronger anesthetics. Interpretations based on hydrophobicity only would require a large change in polarizability between $CFCl_3$—$CHCl_3$, etc. but this is not the case. The data also illustrate the deactivating effect of the fluorines.

With two hydrogens, and CH_2F_2 as the reference molecule, similar observations are made.

All this confirms the $F < Cl < Br$ trend in anesthetic potency (other things being equal) which has been established several years ago by Krantz and Rudo[51)] (Cf. also Larsen[52)] and Clayton[53)]) and the need for involving both polar and non-polar interactions in attempts to explain the mechanism of anesthesia.

Table 4. Polarizabilities and anesthetic potencies of halogen derivatives of ethane

	$\bar{\alpha} \cdot 10^{23}$ cm^{-1}	log 1/p
C_2F_6	(0.398)	
CF_3—CF_2Cl	0.646	~0
CF_2Cl—CF_2Cl	0.838	(0.8–1.1)*
CF_2Cl—$CFCl_2$	1.034	
CF_3—CCl_3	1.047	(1.1–1.6)*
CF_2Br—CF_2Br	1.072	(0.9–1.3)*
CF_2Br—CFClBr	1.252	
CF_3—CHF_2	(0.406)	0.01
CF_3CHCl_2	(0.805)	1.57
CF_3CHClBr (Halothane)	0.937	2.11
CF_2Cl—$CHCl_2$	1.019	
CF_3CH_2F	(0.415)	0.20
CF_3CH_2Cl	(0.615)	0.57
CF_3CH_2Br	0.743	1.00
CF_2Cl—CH_2Cl	0.836	
CF_2Br—CH_2Br	1.063	
CHF_2—CHF_2	(0.415)	
$CHCl_2$—$CHCl_2$	(0.832)	
CHClBr—CHClBr	1.556	
CH_2F—CH_2F	(0.433)	
CH_2Cl—CH_2Cl	0.83	(1.25–1.55)*
CH_2Br—CH_2Br	1.07	(0.98–1.16)*
CHF_2—CF_2OCHF_2	(0.647)	
CHFCl—CF_2OCHF_2 (enflurane)	0.812	
CH_3—OCF_2—CHF_2	(0.664)	
CH_3—OCF_2—CHFCl	0.89	
CH_3O—CF_2—$CHCl_2$ (methoxyflurane)	1.081	2.66

* estimated

Turning to molecules containing two carbon atoms (Table 4) we first consider the group having C_2F_6 as the reference molecule. Gradually replacing F atoms by Cl atoms again yields higher polarizabilities. Replacing them by Br atoms yields even higher ones.

In the next group (CF_3—CHF_2) we have two molecules of high anesthetic potency, CF_3—$CHCl_2$ and CF_3—CHClBr (halothane). If we compare their respective $\bar{\alpha}$ with those of CF_3—CCl_3 we can make the striking observation that the $\bar{\alpha}$ of these strong anesthetics are actually lower than that of CF_3—CCl_3 which has only a very weak anesthetic potency. This again underscores the importance of the acidic hydrogen and that of polar interactions.

Enflurane and methoxyflurane also have moderate $\bar{\alpha}$ values which would not be sufficient to explain their strong anesthetic potency.

On the whole then, these comparisons confirm the general line of thinking of this chapter. The weaker anesthetics act only by dispersion or, possibly, induction forces. Their hydrogen bond perturbing effect is weak and probably indirect through perturbing the hydrophobic parts of the membrane. Molecules like Xe, N_2O, cyclopropane, ethylene are likely to be favored by their small size. The more potent anesthetics act essentially by polar interactions and have a pronounced perturbing effect on hydrogen bonds. Theories should be based on the whole range of intermolecular associations that can occur.

9 Ionization Potentials and Ultraviolet Spectra of Anesthetic Molecules

Ionization potentials play a certain role in interactions through dispersion forces (eq. [1]). They are an important factor too in associations involving charge transfer. While anesthesia does not involve excited states, knowledge of the latter might be useful in assessing the possibility of charge transfer. Therefore, the photoelectron and ultraviolet absorption spectra of a number of anesthetics and similar molecules have been measured and are briefly described here.

The photoelectron spectra were taken on a Perkin-Elmer PS-16[18)] instrument with a HeI source. The ultraviolet absorption spectra were determined in the vapor phase on a McPherson model 225 vacuum ultraviolet spectrometer mounted with a 1200 lines/mm grating and a hydrogen light source. A Cary-17 spectrometer was also used.

According to Koopmans' theorem[54)], the energies of successive photoelectron bands are assigned to ionization from successive molecular orbitals occupied in the electronic ground state so that the band of lowest energy corresponds to the ionization potential with the ion produced in its ground state and the successive bands to production of the ion in an excited state. While this procedure neglects changes in electron correlation and electronic rearrangement occuring when ionization takes place, it can give insight into molecular structure in terms of molecular orbitals at least for valence electrons and is usually readily correlated with the results of quantum chemical calculations. Most discrepancies occur when a molecule possesses close-lying molecular orbitals or when more than one state is issued from a given electron configuration. In such cases the order of photoelectron bands might not be the same as the order of molecular orbitals as it is in the unperturbed ground state.

The photoelectron spectra of halogenated methanes were systematically treated by Turner and his coworkers[55)]. For all Cl, Br and I substituted derivatives the bands of lowest energy correspond to ionization from orbitals of the halogen lone pair type. The respective (vertical) values are, for CH_3Cl, CH_3Br and CH_3I taking the average of the two bands due to spin-orbital splitting: 11.3, 10.7 and 9.8 eV respectively. The orbital is doubly degenerate under C_{3v} symmetry. Conditions are different for fluoromethanes. Due to the high electron attracting power of fluorine the F lone pair electrons require much more energy to be ionized (about

15.5 eV or higher) so that the lowest IP corresponds to orbitals mainly populated in the C—H bonds. It is then natural to find that in methane derivatives containing both fluorine and higher halogens the photoelectron bands of lowest energy are still connected with the lone pairs of the higher halogens but the bands are shifted to higher energies. The IP of CF_3Cl is 13.0 eV, that of CF_3Br and CF_3I are 12.0 and 10.8 eV respectively[56].

In derivatives containing two, three of four higher halogens the lone pair atomic orbitals combine to give 4, 6 or 8 molecular orbitals of the halogen lone pair type some of which are, however, degenerate. Thus, for example, for CH_2Cl_2 there are four such molecular orbitals ($a_1 + a_2 + b_1 + b_2$ under C_{2v} symmetry), for $CHCl_3$ there are four, two of them degenerate ($a_1 + a_2 + e + e$, under C_{3v} symmetry) and for CCl_4 three, all being degenerate ($e + t_1 + t_2$ under T_d symmetry).

The bands of next higher energy in these spectra correspond to orbitals strongly populated in the C—X (X=Cl, Br or I) bonds, followed by those of mainly C—H character. Table 5 lists a few data for simple halogenated methane derivatives relevant to the subsequent discussion on anesthetic molecules[56,58,61].

The photoelectron spectra of ethane derivatives can be interpreted along similar lines. The differences are due mainly to the lower symmetry that makes the degenerate bands split in many cases and to the presence of the C—C bond. The energy of the

Table 5. The ionization potentials (vertical) of a number of halofluorocarbons

Compound	IP (eV)	Reference
CH_3Cl	11.3	(55)
CF_3Cl	13.0	(56) corrected
CF_2HCl	12.6	(56)
CFH_2Cl	11.7	(56)
CH_2Cl_2	11.4	(55)
CF_2Cl_2	12.3	(56)
$CFHCl_2$	12.0	(56)
$CHCl_3$	11.5	(55)
$CFCl_3$	11.9	(56)
CF_3Br	12.0	(56)
CF_2Br_2	11.2	(65)
CH_3CF_2Cl	12.5	(57)
C_2F_5Cl	13.0	(57)
$CF_2Cl—CF_2Cl$	12.85	(57)
$CF_2Br—CF_2Br$	11.4	(61)
$CF_2Cl—CFHCl$	12.0	(57)
$CFCl_2—CF_2Cl$	12.05	(57)
$CF_2Cl—CH_2Cl$	11.7	(59)
$CF_3—CHClBr$ (Halothane)	11.2	(55)
$CH_3OCF_2CHCl_2$ (Methoxyflurane)	11.5	(59)
$CHFCl—CF_2OCHF_2$ (enflurane)	12.2 sh. 12.6	(59)
$CF_3—CCl_3$	11.8	(59)
$CF_3—CHCl_2$	12.0	(59)

orbital mainly populated in the C—C bond is close to that of the orbital mainly populated in the C—Cl and mixing usually occurs. When the molecule contains more than one higher halogen there are again more than one IP related to molecular orbitals based on lone pair type atomic orbitals. Under moderate resolution two Cl atoms in 1,2 position usually give two bands split by about 0.5 eV. Two Cl atoms in 1,1 position generally give four bands spread over a range slightly greater than 1 eV. Table 5 contains several examples.

It is not intended here to give a complete interpretation of these spectra. The bands of lowest energy always relate to molecular orbitals formed from Cl or Br lone pair atomic orbitals. ($\overline{Cl}$, $\overline{Br}$). The first C—Cl bands usually come in near 14 eV or slightly lower so that the assignments in the 13.5–14.0 eV range are somewhat incertain. Above that mixed C—H, C—C and than C—F and $\overline{F}$ ionizations come in. Only the $\overline{Cl}$ and $\overline{Br}$ bands will be treated here, since the bands of higher energy cannot, in any way, be related to the mechanism of anesthesia. Full assignments together with results obtained with a He II source and quantum chemical calculations will be given elsewhere[59].

First we take the methane derivatives. (Table 5) CF_3Cl has a high IP, 13.0 eV. It decreases in the series $CF_3Cl > CF_3Br > CF_3I$. The respective anesthetic potencies increase in the same order. It would be tempting to jump to the conclusion that this indicates charge-transfer interaction. However, these halides could only be electron acceptors in any such interaction, the probable electron donors being oxygen or nitrogen lone pairs, or perhaps bonding π electrons. They all have their IPs much lower, about 9–10 eV. It is possible that the electron affinities follow the same trend as the IPs but these are not known. The IPs vary in opposite direction to the polarizabilities. They might attenuate the dependance of dispersion forces on polarizabilities but the effect is not considered significant. The relative change in the IPs in the series CF_3Cl, CF_3Br, CF_3I is percentagewise moderate. Also the dependence of dispersion energy in $\bar{\alpha}$ is quadratic while the dependance on the IP is linear.

It is more interesting to observe that when a fluorine atom is replaced by a hydrogen, the IP is lower. (CF_3Cl: 13 eV, CF_2HCl: 12.6 eV; CF_2Cl_2: 12.3 eV, $CFHCl_2$: 12.0 eV).

At the same time the lowest ultraviolet absorption frequency also decreases while the anesthetic potency increases. (Table 5). This observation led one of us[19,60] to tentatively involve charge-transfer interactions as a possible factor in the mechanism of anesthesia. While this remains a possibility it has not been proven. It is perhaps interesting to remark, however, that the acidic hydrogen can possibly have an effect in addition to forming hydrogen bonds as a proton donor. "Other things being equal" it lowers the IP while it does not have a significant effect on polarizabilities.

Most of the widely used anesthetics can be considered as ethane derivatives. The photoelectron spectra of a number of fluorochloro-derivatives were published by Chau and McDowell[61] and by Doucet, Sauvageau and Sandorfy[57]. The photoelectron and ultraviolet spectra of the most widely used anesthetics like halothane, methoxyflurane and enflurane as well as those of $CF_2Cl—CF_2Cl$, $CF_3—CCl_3$ and $CF_3—CHCl_2$ have been measured by us and will be briefly described.

The IP of C_2F_5Cl is 12.96 eV[57], practically the same as for CF_3Cl. If a chlorine is substituted on both carbons, ($CF_2Cl—CF_2Cl$) the band splits into four. At sufficiently high resolution they are found at 12.47, 12.82, 13.06 and 13.19 eV[61]. If a hydrogen is introduced ($CF_2Cl—CFHCl$) there is a shift to lower energies. Only two bands could be resolved, centered at 12.0 and 12.6 eV[57]. This is similar to the observation made for methane derivatives.

For $CFCl_2—CF_2Cl$ four bands were resolved, at 12,0, 12,5. 13.0 and 13.2 eV. The lower ones can be quite safely assigned to the $CFCl_2$ part. $CF_2Cl—CH_2Cl$ has bands at 11.7, 11.8 and 12.6 eV of which the two of lower energy can only correlate to the CH_2Cl part and 12.6 to the CF_2Cl part. The split in the former might be due to the loss of threefold symmetry or to partly resolved vibrational fine structure.

The spectrum of $CF_2Br—CF_2Br$ has been determined by Chau and McDowell[61]. The lone pair bands are at 11.44, 11.83, 12.11 and 12.2 eV, roughly 1 eV lower than for $CF_2Cl—CF_2Cl$. The replacement of one of the fluorine atoms by hydrogen, as expected, lowers them further: 10.86, 11.14, 11.46 and 11.65 eV. The lowest of these quite certainly belong to the CFHBr part.

The comparison between $CF_3—CCl_3$ and $CF_3—CHCl_2$ is interesting. The difference between the two consists in the replacement of a *chlorine* by hydrogen. It does not alter the spectrum profoundly although an additional band is resolved for $CF_3—CCl_3$ at 11.8 eV. The band centers are at 11.8, 12.2 and 13.2 eV for $CF_3—CCl_3$ and 12.0 and 12.7 (a double band) for $CF_3—CHCl_2$. However, a shoulder at the low energy side of the 12.0 eV peak of the latter shows that the lowest IPs have about the same value for both compounds. Now, while $CF_3—CHCl_2$ has a high anesthetic potency that of $CF_3—CCl_3$ is weaker. This seems to substantiate the contention that the important factor is the proton donor property of the hydrogen containing molecule, not its effect on the IP.

Halothane (whose photoelectron spectrum was measured previously by Turner et al. ([55], p. 242) has the $\overline{Br}$ bands at 11.24 and 11.45, the $\overline{Cl}$ bands at 12.20 and 12.32 eV showing that the interaction between the $\overline{Br}$ and $\overline{Cl}$ lone pairs is not strong. This is another case of a molecule having both an acidic hydrogen and a low IP and a high anesthetic potency.

The case of methoxyflurane ($CH_3OCF_2CHCl_2$) is similar (Fig. 9). The four $\overline{Cl}$ IPs are resolved at 11.5, 12.2, 12.4 and 13.2 eV. Comparison with the spectrum of $CF_3—CHCl_2$ shows that the methoxy group has a lowering effect on these IPs. It is of some interest to search for the photoelectron band which corresponds to ionization from the essentially oxygen lone pair orbital ($\overline{O}$). Hardin and Sandorfy[49] identified this band near 14 eV in fluoroethers and there is no reason to expect it at much lower energies for methoxyflurane. This is even more so for enflurane in which both neighbours of the oxygen atom are fluorinated.

The lowest bands of enflurane (CF_2HOCF_2CHFCl) are at 12.2 (shoulder) and 12.6 eV. Clearly they correlate to the chlorine lone pair whose energy is split due to the lack of symmetry in this molecule (Fig. 10).

All these data are contained in Table 5. We now turn our attention to the ultraviolet absorption spectra. As is well known[56,57,58,62,63] the lowest singlet-singlet absorption band of alkyl chlorides, bromides and iodides is due to a valence-shell type transition from an essentially halogen lone pair orbital ($\overline{X}$) to an

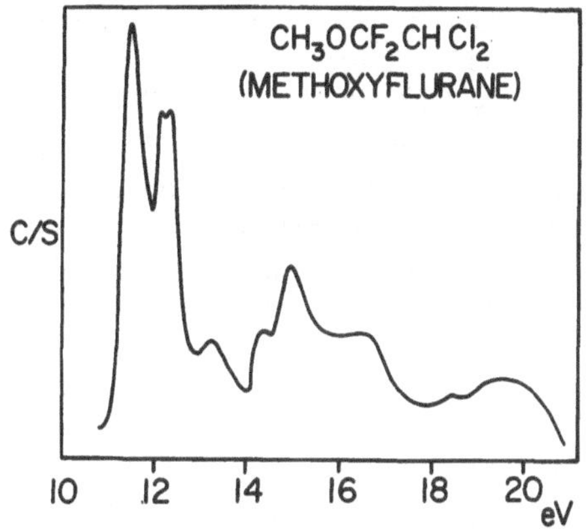

Fig. 9. The HeI photoelectron spectrum of methoxyflurane, $CH_3OCF_2CHCl_2$

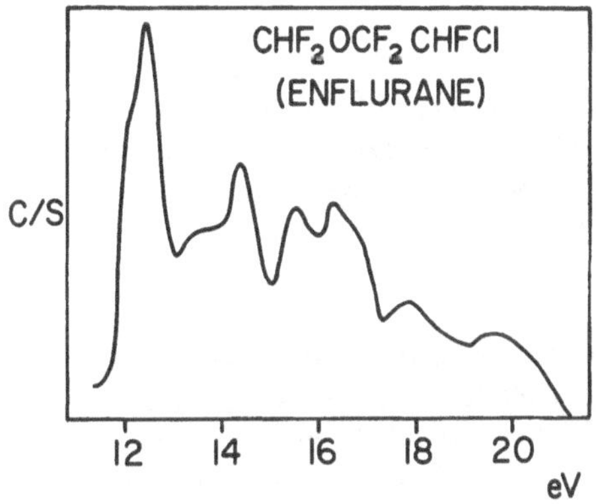

Fig. 10. The HeI photoelectron spectrum of enflurane, CHF_2OCF_2CHFCl

orbital which is antibonding in the C—X bond: $\bar{X} \rightarrow (C—X)^*$. The same is true for chloro, bromo and iodo fluorocarbons. The effect of the fluorines is generally a shift to higher frequencies. For the sake of example, we can give the respective band centers for CF_3Cl, CF_3Br and CF_3I which are at 65000, 48850[46] and 37700 cm^{-1} [64] respectively. When the molecule contains more than one of the same halogen (other than fluorine) the bands split as do the ionization potentials. Since these bands are broad, however, it is usually not possible to identify all the bands resulting from the splitting. For example CF_2Cl_2 has two broad bands in this spectral region centered at 56460 and 65400 cm^{-1} respectively. For $CFCl_3$ the $\overline{Cl} \rightarrow (C—Cl)^*$ bands are found between 54000 and 60500 cm^{-1}. Conditions are similar for fluoroethane derivatives. For C_2F_5Cl the band is at 66700 cm^{-1}, for $CF_2Cl—CF_2Cl$ at 64500 cm^{-1} and for $CFCl_2—CF_2Cl$ at 62500 cm^{-1}. In general the highest frequencies belong to the molecules having only $—CF_2Cl$ groups, $—CFCl_2$ groups lead to absorption at lower frequencies and $—CCl_3$ lower still. At higher frequencies we find strong $\bar{X}$ to Rydberg bands in all these spectra but these will not concern us here. Instead we refer to the original publications[56,57].

CF_3—$CHCl_2$ has its lowest $\overline{Cl} \rightarrow (C—Cl)^*$ band at 57100 cm^{-1} with a molar extinction coefficient $\varepsilon_{max} \cong 520$ liters M^{-1} cm^{-1}. More intense bands follow at 69500 ($\varepsilon_{max} \cong 4500$) and 74000 cm^{-1} ($\varepsilon_{max} \cong 9500$) the latter exhibiting two well pronounced shoulders. Since the IP of this molecule is at about 95000 cm^{-1} the term value for the 57100 cm^{-1} band would be about 38000 cm^{-1}, much too high for any Rydberg band. This is in line with the anticipated assignment to a $\overline{Cl} \rightarrow (C—Cl)^*$ valence-shell type transition. The band at 69500 would fit a $\overline{Cl} \rightarrow 3s$ Rydberg assignment but it could also correspond to a second valence-shell transition from the higher $\overline{Cl}$ orbitals. The large band at 74000 cm^{-1} can receive contributions from a variety of 3s and 3p type transitions originating from one of the $\overline{Cl}$ orbitals relating to the four $\overline{Cl}$ IPs. In view of the diffuse character of these spectra it would be vain to make more precise assignments.

The spectrum of CF_3—CCl_3 is of the same general type; the band of lowest frequency is centered at 57500 cm^{-1} ($\varepsilon_{max} \sim 1080$).

The spectrum of halothane (CF_3—CHClBr) is shown in Fig. 11. The presence of the bromine atom gives rise to a $\overline{Br} \rightarrow (C—Br)^*$ type valence-shell transition. Its center is near 49000 cm^{-1} ($\varepsilon_{max} \sim 340$). The band of next higher frequency 59500 ($\varepsilon_{max} \sim 2500$) is readily assigned to the $\overline{Cl} \rightarrow (C—Cl)^*$ transition. It obviously receives its intensity by overlap with the wing of the much more intense band that follows at higher frequencies. The diffuse band at about 66000 cm^{-1} has the right term value (about 24000 cm^{-1}, the $\overline{Br}$ IP being about 90000 cm^{-1}) to be $\overline{Br} \rightarrow 3s$, the band at 70700 cm^{-1} to be $\overline{Cl} \rightarrow 3s$. In the same way the bands at 72500 cm^{-1} can be assigned to $\overline{Br} \rightarrow 3p$ (term value 17500 cm^{-1}), the bands that follow at higher frequencies to $\overline{Cl} \rightarrow 3p$ and $Br \rightarrow 3d$, $\overline{Cl} \rightarrow 3p$ etc. The high intensities of these bands could indicate some valence-shell admixture.

As expected methoxyflurance ($CH_3OCF_2CHCl_2$) (Fig. 12) has a spectrum very similar to that of CF_3—$CHCl_2$ with the first band at 57500 cm^{-1}. Indeed, for both molecules the transitions originate with the $\overline{Cl}$ levels and the methoxy group can cause only slight shifts of the band frequencies. The absence of any bands originating with the oxygen lone pair confirms our assigment of the $\overline{O}$ IP to a high energy photoelectron band (about 14 eV).

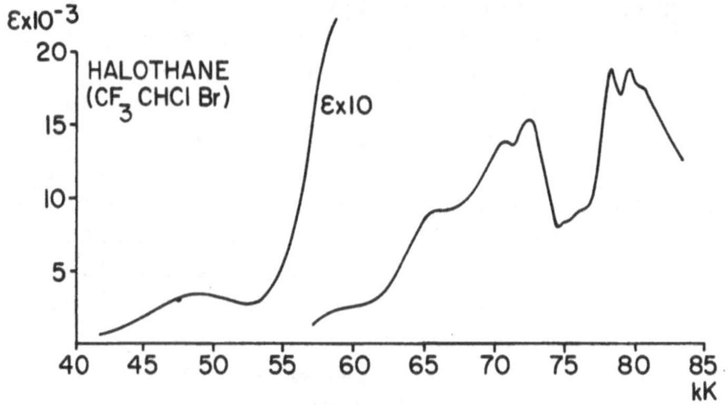

Fig. 11. The far ultraviolet absorption spectrum of halothane, CF_3—CHClBr

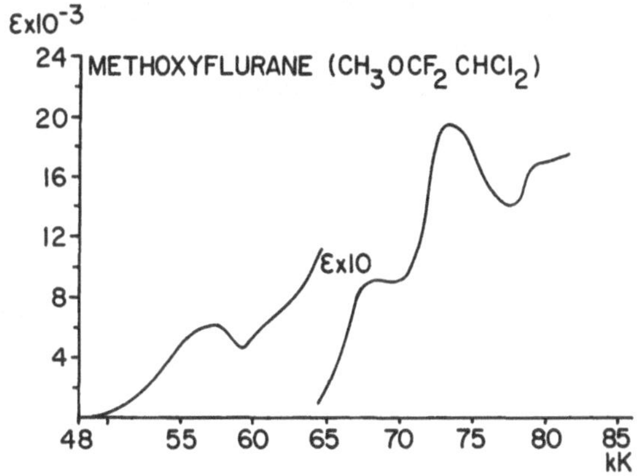

Fig. 12. The far ultraviolet absorption spectrum of methoxyflurane, $CH_3OCF_2CHCl_2$

Enflurane (CF_2HOCF_2CHFCl) contains only one Cl and it is attached to a carbon to which an H and an F atom are also attached. The first band is a shoulder at about 57500 cm^{-1}, followed by another shoulder near 60000 and a peak at 64000 cm^{-1}. The complexity of the spectrum is probably due to the significant split in the $\overline{Cl}$ IPs, about 0.4 eV or 3200 cm^{-1}. The more intense Rydberg bands are located near 70500, 74700 and 78000 cm^{-1}.

Concluding this section all that one can say is that we found no relationship between anesthetic potency and either the ionization potentials or the frequency of the lowest ultraviolet absorption band. The observation that replacement of a fluorine atom by a hydrogen usually lowers the IP is probably of some value. However, as was pointed out above this could only indicate the possibility of charge transfer interaction if the electron affinities followed the same trend. Unfortunately these have not been determined and the variations in the frequencies of the broad UV bands are too irregular to draw conclusions. It seems that there exists an indirect relationship between the acidity of these molecules and their IPs and what counts is their proton donor ability connected with the "acidic hydrogen" as has been concluded from the infrared studies described in previous sections.

10 Concluding Remarks

The results presented in this chapter were based on spectroscopic investigations on model systems. Similar experiments on truly biological material remain to be performed. However, the model systems that have been chosen contain some of the most important patterns of molecular association that determine the functioning of the nervous system and many other phenomena connected with the living organism. Anesthesia, a relatively light perturbation of the nervous system appears to open the way into the realm of these phenomena. We have to move much closer to the nerve cell and its all important membrane. It is believed, however, that model studies might lead the researcher in his endeavor to gain knowledge at the molecular level.

Our main conclusion at this stage is that the full scale of molecular associations, polar and non polar, from hydrogen bonds to hydrophobic interactions has to be invoked if we are to understand anesthesia and the many other phenomena which in the living organism depend on molecular associations.

11 References

1. Seeman, P.: Pharmacol. Rev. *24*, 583 (1972)
2. Miller, K. W., Smith, E. B.: in A Guide to Molecular Pharmacology — Toxicology. Part 2 (Ed. R. M. Featherstone), pp. 427–475, M. Dekker, New York, 1973
3. Halsey, M. J.: in Anesthetic Uptake and Action. (Ed. E. I. Eger), pp. 45–76. Williams and Wilkins, Baltimore, 1974
4. Miller, J. C., Miller, K. W.: in Physiological and Pharmacological Biochemistry (Ed. H. K. F. Blaschko), pp. 33—75, Butterworths, London, 1975
5. Kaufman, R. D.: Anesthesiology *46*, 49 (1977)
6. Denson, D. D.: Chemtech. *7*, 446 (1977)
7. Changeux, J. P. et al.: in Molecular Properties of Drug Receptors (Ed. R. Porter and M. O'Connor), pp. 197, Churchill, London, 1970
8. Eyring, H., Woodbury, J. W., D'Arrigo, J. S.: Anesthesiology *38*, 415 (1973)
9. Urry, D. W. et al. in press
10. Davies, M. M.: Some Electrical and Optical Aspects of Molecular Behaviour, Chapter 7, Pergamon Press, Oxford, 1965
11. Pryde, J. A.: The Liquid State, Chapter 4, Hutchinson, London, 1966
12. Meyer, H.: Naunyn-Schmiedebergs Arch. Exp. Pathol. Pharmacol. *42*, 109 (1899)
13. Overton, E.: Studien über die Narkose zugleich ein Beitrag zur allgemeinen Pharmacologie, Gustav Fischer, Jena, 1901
14. Hansch, C. et al.: J. Med. Chem. *18*, 546 (1975)
15. Di Paolo, T., Kier, L. B., Hall, L. H.: Mol. Pharmacol. *13*, 31 (1977)
16. Nemethy, G., Scheraga, H. A.: J. Chem. Phys. *36*, 3382 (1962)
17. Bernard-Houplain, M. C., Sandorfy, C.: J. Chem. Phys. *56*, 3412 (1972)
18. Bernard-Houplain, M. C. et al.: Chem. Phys. Lett. *11*, 149 (1971)
19. Di Paolo, T., Sandorfy, C.: J. Med. Chem. *17*, 809 (1974)
20. Di Paolo, T., Sandorfy, C.: Can. J. Chem. *52*, 3612 (1974)
21. Bicca de Alencastro, R., Sandorfy, C.: Can. J. Chem. *50*, 3594 (1972)
22. Massuda, R., Sandorfy, C.: Can J. Chem. *55*, 3211 (1977)
23. Nagyrevi, A., Sandorfy, C.: Can. J. Chem. *55*, 1592 (1977)
24. Davies, R. H., Bagnall, R. D., Jones, W. G. M.: Int. J. Quantum Chem. Quantum Biol. Symp. *1*, 201 (1974)
25. Davies, R. H. et al.: ibid. *3*, 171 (1976)
26. Brown, J. M., Chaloner, P. A.: Can. J. Chem. *55*, 3380 (1977)
27. Koehler, L. S., Curley, W., Koehler, K. A.: Mol. Pharmacol. *13*, 113 (1977)
28. Sandorfy, C.: Anesthesiology *48*, 357 (1978)
29. Whetsel, K. B.: Spectrochim. Acta. *17*, 614 (1961)
30. Allerhand, A., Schleyer, P. R.: J. Am. Chem. Soc. *85*, 371, 1233 (1963)
31. Stevenson, D. P., Coppinger, G. M.: ibid. *84*, 149 (1962)
32. Gomel, M.: Ann. Chim. *3*, 415 (1968)
33. Bellamy, L. J., Pace, R. J.: Spectrochim. Acta. *22*, 535 (1966)
34. Suckling, C. W.: Brit. J. Anaesth. *29*, 466 (1957)
35. Martire, D. E. et al.: J. Am. Chem. Soc. *98*, 3101 (1976)
36. Luck, W. A. P., Ditter, W.: J. Mol. Struct. *1*, 261, 339 (1967–1968)
37. Burneau, A., Corset, J.: J. Chim. Phys. *69*, 171 (1972)
38. Bourdéron, C., Sandorfy, C.: J. Chem. Phys. *59*, 2527, (1973)
39. Di Paolo, T., Bourdéron, C., Sandorfy, C.: Can. J. Chem. *50*, 3161 (1972)
40. Trudeau, G. et al.: Can. J. Chem. *56*, 1681 (1978)
41. Trudeau, G.: M. Sc. Thesis. Université de Montréal 1979

42. Kanbayashi, U.: Bull. Chem. Soc. Japan *36*, 1173 (1963)
43. Evans, J. C., Lo, Y. S.: Spectrochim. Acta. *21*, 33 (1965)
44. Oi, N., Coetzee, J. F.: J. Am. Chem. Soc. *91*, 2473 (1969)
45. Green, R. D.: Hydrogen Bonding by CH Groups, pp. 49, Wiley, New York, 1974
46. Gasanov, R. G.: Optics and Spectroscopy, Engl. Ed., *23*, 294 (1967)
47. Herzberg, G.: Molecular Spectra and Molecular Structure, Vol. II, p. 316, Van Nostrand, New York, 1945
48. Nielsen, J. R., Liang, C. Y.: J. Chem. Phys. *21*, 1069 (1953)
49. Hardin, A. H., Sandorfy, C.: J. Fluor. Chem. *5*, 435 (1975)
50. Vogel, A. I. et al.: J. Chem. Soc. 514 (1952)
51. Krantz, J. C., Rudo, F. G.: Handbook of Experimental Pharmacology. Vol. XX/1, pp. 501–564, (Ed. F. A. Smith), Springer, Berlin 1966
52. Larsen, E. R.: Fluorine Chem. Rev. *3*, 1 (1969)
53. Clayton, J. W.: ibid. *1*, 197 (1967)
54. Koopmans, T.: Physica *1*, 104 (1934)
55. Turner, D. W. et al.: Molecular Photoelectron Spectroscopy, Wiley-Interscience, London, 1970
56. Doucet, J., Sauvageau, P., Sandorfy, C.: J. Chem. Phys. *58*, 3708 (1973)
57. Doucet, J., Sauvageau, P., Sandorfy, C.: ibid. *62*, 355 (1975)
58. Sandorfy, C.: Atm. Env. *10*, 343 (1976)
59. Dumas, J. M. et al. to be published
60. Sandorfy, C.: in Environmental Effects on Molecular Structure and Properties, p. 529, (Ed. B. Pullman) Reidel, Dordrecht, Holland, 1976
61. Chau, F. T., McDowell, C. A.: J. Phys. Chem. *80*, 2923 (1976)
62. Zobel, C. R., Duncan, A. B. F.: J. Am. Chem. Soc. *77*, 2611 (1955)
63. Russell, B. R., Edwards, L. O., Raymonda, J. W.: ibid. *95*, 2129 (1973)
64. Dunn, T. M., Herzberg, G.: Quoted in G. Herzberg, Molecular Spectra and Molecular Structure, Vol. III, p. 532, Van Nostrand, Princeton, New Jersey, 1966
65. Doucet, J. et al.: J. Chem. Phys. *62*, 366 (1975)
66. Sandorfy, C., Trudeau, G.: in Infrared and Raman Spectroscopy of Biological Molecules, p. 319, (Ed. Th. Theophanides), Reidel, Dordrecht, Holland, 1979

MIX
Papier | Fördert gute Waldnutzung
FSC® C083411

Zeitfracht Medien GmbH
Ferdinand-Jühlke-Straße 7
99095 Erfurt, Deutschland
produktsicherheit@kolibri360.de